Intelligent Moving Cities: Technological Leap and Social Integration of Autonomous Mobile Robots

Minje Choi · Seungjae Lee

Intelligent Moving Cities: Technological Leap and Social Integration of Autonomous Mobile Robots

Springer

Minje Choi
Department of Land Economy
University of Cambridge
Cambridge, UK

Seungjae Lee
Department of Transportation Engineering
University of Seoul
Dongdaemun, Korea (Republic of)

ISBN 978-981-96-8863-0 ISBN 978-981-96-8864-7 (eBook)
https://doi.org/10.1007/978-981-96-8864-7

This Springer imprint is published by the registered company Springer Nature Singapore Pte Ltd.
The registered company address is: 152 Beach Road, #21-01/04 Gateway East, Singapore 189721, Singapore

Preface

Cities have continuously evolved and developed. From the Industrial Revolution to the Information Revolution, we are now entering another era of innovation led by artificial intelligence and autonomous mobile robots (Auto-Mobile Robots). This book, Intelligent Moving Cities: Technological Leap and Social Integration of Autonomous Mobile Robots, focuses on exploring how technology and society can coexist harmoniously in the midst of these changes.

Autonomous mobile robots have the potential to fundamentally alter not only transportation but also the structure and function of cities, as well as the daily lives of citizens. When combined with artificial intelligence, autonomous driving technology affects a wide range of areas, including logistics, public transportation, pedestrian environments, and public services. These impacts go beyond mere technological advancements and are becoming critical elements in building human-centered smart cities.

However, these changes cannot be viewed solely from a technological perspective. We must ensure that new mobility technologies do not exacerbate social inequalities. Will technological innovation benefit only large corporations and specific social groups while creating another form of disparity for vulnerable populations? While autonomous mobile robots and AI-based transportation systems are expected to make cities more efficient, discussions are also needed on changes in the labor market, data privacy issues, and potential conflicts between pedestrians and vehicles.

This book emphasizes the necessity of balancing technological innovation with social acceptance. For autonomous mobile robots to operate smoothly in real urban environments, infrastructure modifications and legal and ethical considerations must also be addressed. How should we design cities where robots and humans coexist, and what roles should policymakers, businesses, researchers, and citizens play? This book aims to provide insights into these critical questions.

In particular, we must also consider how to balance public interest with individual freedom within cities. For instance, even if autonomous mobile robots optimize transportation and logistics systems, will this come at the cost of restricting pedestrian mobility? Will road infrastructure have to be excessively modified to accommodate specific technologies? Additionally, if robots and AI replace human labor in cities, are adequate social protections being put in place to address this shift?

This book is not merely about technological advancements but serves as a platform for discussion on ensuring that autonomous mobile robots and smart cities develop in a way that remains human-centered. Technology should be a tool to assist humanity, not a force that subjugates it. With this awareness, we analyze how autonomous mobile robots will impact citizens' lives and what policy responses must be prepared.

The authors aim to convey that autonomous mobile robots are more than just technological advancements—they are pivotal factors driving social, economic, and environmental changes. As technological progress accelerates, understanding and preparation must be thorough.

We hope this book will provide readers with an opportunity to deeply explore the possibilities of the new cities led by autonomous mobile robots, engage in discussions, and contemplate the direction for future progress.

Cambridge, UK Minje Choi
Dongdaemun, Korea (Republic of) Seungjae Lee

Competing Interests The authors declare that they have no competing interests associated with this study.

About This Book

Intelligent Moving Cities: Technological Leap and Social Integration of Autonomous Mobile Robots explores the role of artificial intelligence (AI) and autonomous mobile robots in modern cities. With rapid technological advancements and the expansion of smart city infrastructure, we must consider how autonomous mobile robots impact existing transportation systems, logistics networks, and citizens' daily lives. This book analyzes these changes, their social, ethical, and policy implications and how autonomous mobile robots can develop in harmony with environmental sustainability in urban settings.

This book focuses on maintaining a balance between technological advancement and social integration. It emphasizes that technological innovation should not only enhance efficiency but also uphold a human-centered approach. It also provides insights for researchers, policymakers, entrepreneurs, and citizens on how to effectively utilize autonomous mobile robots.

Structure of the Book

This book consists of eight main chapters, each addressing a specific aspect of autonomous mobile robots and smart cities.

Chapter 1: Systematic Analysis of the Need for Autonomous Robots.

This chapter systematically analyzes why autonomous mobile robots are necessary in urban environments. It explains the limitations of existing transportation and logistics systems and discusses the problems that autonomous mobility technologies can solve and the expected benefits. It highlights the historical context and the integration of these technologies with existing urban infrastructure, emphasizing their significance in modern society.

Chapter 2: Advances in Autonomous Driving Robots: Past, Present and Future.

This chapter describes the core technologies of autonomous mobile robots, including sensors, AI, and computer vision, and explores their real-world applications. It examines their potential use in logistics, public transportation, and industrial automation, comparing current technological capabilities with future development possibilities.

Chapter 3: Technical Considerations for Autonomous Navigation and Route Navigation.

This chapter covers the key technologies required for the effective operation of autonomous mobile robots, such as navigation and path-planning algorithms. It explains how robots adapt to dynamic urban environments and find optimal routes in real time. Additionally, it analyzes how dynamic traffic flow and predictive models enhance the efficiency of robotic movement.

Chapter 4: Hardware Design and Scalability of Autonomous Robots.

This chapter analyzes the principles of hardware design for autonomous mobile robots, focusing on durability, safety, and scalability. It discusses strategies to ensure that robots remain functional in urban settings for extended periods. It also covers material selection, energy efficiency, and maintenance convenience.

Chapter 5: Various Applications of Autonomous Robots.

This chapter explores how autonomous mobile robots are being used in public safety, personal and public transportation, and industrial and commercial sectors. It examines real-world implementations and discusses their future development potential. Additionally, it highlights how robots improve pedestrian environments and assist people with disabilities.

Chapter 6: Integration of Autonomous Robots in Environmental and Urban Planning.

This chapter analyzes how autonomous mobile robots integrate with urban infrastructure and contribute to smart city strategies. It discusses environmental sustainability and policy considerations, exploring how these robots impact society and how sustainable urban transportation policies can be implemented.

Chapter 7: Challenges and Prospects for Autonomous Robots.

This chapter examines the social, ethical, and legal challenges associated with autonomous mobile robots. It discusses how public policies, regulations, and data privacy issues affect technological acceptance and proposes solutions. It also explores the relationship between autonomous mobile robots and labor markets, addressing the impact on jobs and the need for workforce retraining.

Chapter 8: Environmental Considerations and Sustainability of Autonomous Robots.

This chapter assesses the environmental impact of designing, developing, and operating autonomous mobile robots and proposes sustainability-driven solutions. It discusses lifecycle assessments (LCA), the use of eco-friendly materials, and energy efficiency improvements. Additionally, it examines strategies for reducing carbon emissions and managing smart energy systems to ensure environmentally responsible robotic operations.

Appendix: Automated Pedestrian Environment Assessment Using Autonomous Robots.

This appendix describes an innovative approach for assessing pedestrian environments using autonomous robots and deep learning techniques.

Traditional sidewalk assessments often rely on subjective judgments, causing inconsistent and unreliable results. Manual inspections require significant labor and are prone to human bias, highlighting the need for a standardized, objective evaluation method.

As urban planning increasingly prioritizes pedestrian-friendly environments, accurately assessing sidewalk conditions is essential for accessibility and safety. To address these issues, this chapter introduces a novel method utilizing autonomous robots and deep learning-based image analysis. By defining standardized assessment criteria, this approach ensures consistent, efficient, and objective pedestrian environment evaluations, supporting informed urban planning and resource management.

Contents

1 Systematic Analysis of the Need for Autonomous Robots 1
1.1 Introduction 1
1.2 The Intersection of Technological Progress and Social Demand 4
1.2.1 Surroundings Recognition Technology 6
1.2.2 Artificial Intelligence 7
1.2.3 Computing Technology 9
1.2.4 Communication and Network Technology 11
1.2.5 Autonomous Driving Applications 13
1.2.6 Demand for Autonomous Driving 14
1.3 Historical Context and Evolution of Autonomous Driving Technology 18
1.4 Expected Effects and Role of Autonomous Robots in the Future Society 20
References 22

2 Advances in Autonomous Driving Robots: Past, Present and Future 25
2.1 The Evolution of Autonomous Driving: From Past to Present 26
2.1.1 The Birth of Autonomous Driving: Early Developments (1980s–1990s) 26
2.1.2 The Rise of AI and Sensor Fusion (2000s–2010s) 27
2.1.3 Current Technology Landscape: Advanced Models and Key Milestones (2010s–2020s) 29
2.2 Current Applications and Achievements of Autonomous Driving Technology 30
2.2.1 Autonomous Driving in Urban Environments 30
2.2.2 Industrial and Specialty Applications 31
2.2.3 Autonomous Driving in Public Services 32

2.3 Future Prospects and Developments in Autonomous Driving Technology 32
2.3.1 Technological Innovations: AI, Edge Computing, and Beyond 33
2.3.2 Application Expansion and City Integration 34
2.3.3 Social and Ethical Considerations 34
2.3.4 Future Challenges and Opportunities 35
References 36

3 Technical Considerations for Autonomous Navigation and Route Navigation 39
3.1 The Basics of How to Build a Navigation System 40
3.1.1 Overview of Navigation and Route Finding Algorithms 40
3.1.2 Comparison of Global and Local Route Planning 42
3.1.3 Key Technology for Supporting Path Searching 43
3.2 Continuous Development of D* Algorithm Under Autonomous Mobile Robots 44
3.2.1 D* Algorithm and Application 44
3.2.2 Modifications to D* for Pedestrian Interaction 45
3.2.3 Challenges and Future Directions 47
3.3 Real-Time Route Planning and Energy Efficiency Considerations 48
3.3.1 Real-Time Route Planning in Dynamic Environments 48
3.3.2 Energy Efficiency of Autonomous Mobile Robots 49
3.3.3 Balancing Real-Time Responsiveness and Energy Efficiency 50
References 51

4 Hardware Design and Scalability of Autonomous Robots 53
4.1 Core Hardware Components and Functional Requirements 54
4.1.1 Essential Components Overview 54
4.1.2 Balance of Function and Aesthetic Sense 55
4.1.3 Human-Centered Ergonomics 56
4.2 Modular Design and Customized Scalability 57
4.2.1 Modular Hardware Architecture 57
4.2.2 Adaptability to User Needs 59
4.2.3 Scalability for Urban Dynamics 59
4.3 Durability and Safety Design Considerations 61
4.3.1 Robustness in a Variety of Environments 61
4.3.2 Safety Features for Interaction with People 62
4.3.3 Testing and Certification Standards 62
References 63

5 Various Applications of Autonomous Robots 65
5.1 Utilization of Robots for Public Safety and Security 66
5.1.1 Patrol and Surveillance Robots 67
5.1.2 Disaster Response and Rescue Robots 67
5.2 Cleaning and Maintenance Robots 69
5.3 Personal and Public Mobility Solutions 72
5.3.1 Personal Mobility Robots 72
5.3.2 Freight Transport Assistance 72
5.3.3 Integration with Public Transportation 74
5.4 Commercial and Industrial Sectors 75
5.4.1 Logistics and Delivery Robots 75
5.4.2 Agricultural and Construction Robots 76
5.4.3 Manufacturing and Maintenance 78
References 80

6 Integration of Autonomous Robots in Environmental and Urban Planning 83
6.1 Introduction of Autonomous Robots into Urban Environments and Their Impact 84
6.1.1 The Significance of Introducing Autonomous Robots into Urban Areas 85
6.1.2 Enhancement of Urban Quality of Life 86
6.2 Interaction Between Public Transportation and Autonomous Robots 88
6.2.1 Role of Autonomous Robots in Promoting Public Transportation 88
6.2.2 Mobility Hubs and Autonomous Systems 89
6.2.3 Enhancing User Experience and Utility 92
6.3 Strategies for Integrating Urban Planning and Autonomous Robots 93
6.3.1 Policy Consensus and Planning 93
6.3.2 Industrial Potential of Autonomous Robots 94
6.3.3 Integration with Urban Planning 95
References 97

7 Challenges and Prospects for Autonomous Robots 99
7.1 The Technical and Ethical Challenges 100
7.1.1 The Ethical Dilemma of Autonomous Decision-Making 100
7.1.2 Protection of Personal Information and Data 101
7.1.3 Trust and Public Perception 102
7.2 The Regulatory Framework and Legal Considerations 103
7.2.1 Development of Regulatory Standards 103
7.2.2 Liability and Legal Responsibility 104
7.2.3 Synergy of Ethics and Regulation 105

7.3 The Long-Term Perspective and Social Acceptance 106
7.3.1 The Role of Public Education and Participation 106
7.3.2 Long-Term Vision for Social Integration of Autonomous Robots 107
7.3.3 Building a Framework for Social Acceptance 108
References 111

8 Environmental Considerations and Sustainability of Autonomous Robots 113
8.1 Reusable Hardware and Circular Design 115
8.1.1 Inspiration from Aerospace: Lessons from Space-X and Beyond 115
8.1.2 Designing Robots with Lifespan and Scalability in Mind: Circular Economy 116
8.2 Carbon Neutrality and Lifecycle Assessment 118
8.2.1 Carbon Footprint of Robot Manufacturing 118
8.2.2 Carbon Neutral Operation: Renewable Energy and Efficiency 120
8.2.3 Lifecycle Assessment for Sustainable Robots 121
8.3 Environmental Impact of Backend Support Systems 122
8.3.1 Energy Use in Data Processing and Cloud Support 122
8.3.2 Green Data Centers and Sustainable Infrastructure 124
8.3.3 Autonomous Robots as Environmental Stewards 127
References 129

Appendix: Automated Pedestrian Environment Assessment Using Autonomous Robots 131

List of Figures

Fig. 1.1	Changes in transportation	2
Fig. 1.2	Technologies in autonomous robots	5
Fig. 1.3	Advances in surrounding recognition technology	7
Fig. 1.4	Mark I, the world's first large-scale digital computer (Getty Images by PhotoQuest)	10
Fig. 1.5	Evolution of telecom networks from 1 to 6G	12
Fig. 1.6	Older population as a percentage of the labor force (*Source* OECD)	16
Fig. 1.7	Layers in a software-define vehicles (*Source* AUTOCRYPT)	16
Fig. 1.8	Applications of autonomous robots	17
Fig. 1.9	Starship autonomous delivery robots (*Source* Starship technology)	18
Fig. 1.10	Autonomous outdoor security robot in operation in the Times Square subway (*Source* Jefferson Siegel for The New York Times)	18
Fig. 1.11	Sixty-five years of automotive steady stages (*Source* Rosenzweig & Bartl, 2015)	19
Fig. 2.1	The grand challenge (*Source*: https://www.darpa.mil/about/innovation-timeline)	27
Fig. 2.2	DARPA urban challenge (*Source* https://www.darpa.mil/about/innovation-timeline)	28
Fig. 2.3	Autonomous technology: 3D point clouds captured by LiDAR (*Source* Liu et al., 2023)	28
Fig. 2.4	Autonomous shuttles (*Source* PPPR, 2020)	30
Fig. 2.5	Nuro's driverless delivery service (*Source* Das, 2021)	31

Fig. 2.6 Alibaba's autonomous robots in Warehouse (*Source* Prawira, 2023) ... 32
Fig. 2.7 Edge computing-based connected autonomous Vehicles (*Source* Gupta et al., 2023) ... 33
Fig. 2.8 Architecture for connected autonomous vehicles in MaaS (*Source* Lee & Madnick, 2021) ... 34
Fig. 2.9 Smart cities communications architecture with autonomous vehicles (*Source* Talib et al., 2018) ... 36
Fig. 3.1 Illustration of Dijkstra's algorithm (*Source* Rehman et al., 2019) ... 41
Fig. 3.2 Illustration of algorithm: **a** A* algorithm, **b** D* algorithm (*Source* Kim et al., 2018) ... 41
Fig. 3.3 D* algorithm diagram (*Source* Jin et al., 2023) ... 42
Fig. 3.4 **a** Turtlebot's LiDAR and camera-based data visualization, **b** Turtlebot ... 43
Fig. 3.5 Avoidance simulation of two obstacles (*Source* Jung et al., 2010) ... 45
Fig. 3.6 Finding the average avoidance point to prevent collisions with pedestrians (Choi et al., 2024a) ... 46
Fig. 3.7 Modified D* algorithm driving behavior: **a** $D \geq 1.77$ m, **b** $D < 1.77$ m (Choi et al., 2024a) ... 47
Fig. 3.8 The process of environmental recognition using multiple sensor ... 48
Fig. 3.9 Example of an energy-saving scenario by adjusting the LiDAR detection range (Choi et al., 2024b) ... 50
Fig. 3.10 Graph of Lidar range and power consumption analysis (Choi et al., 2024b) ... 51
Fig. 4.1 The components of the robot ... 55
Fig. 4.2 Example of a robot that fits the user's characteristics: Porter robot ... 56
Fig. 4.3 Modular hardware (*Source* Korean Agency for Technology and Standards) ... 58
Fig. 4.4 Example of using a module in a PBV (*Source* Hyundai Motor Group) ... 58
Fig. 4.5 Low density suburbs and high density city center ... 60
Fig. 4.6 Autonomous robots in various conditions ... 61
Fig. 4.7 Examples of field testing sites (*Source* UCL PEARL Center) ... 63

Fig. 5.1 Knightscope robot for patrol and surveillance (*Source* https://www.knightscope.com/) 68
Fig. 5.2 Overseas disaster response robots (*Source* https://web.archive.org/web/20080808120417/http://www.fostermiller.com/lemming.htm) by TALON, PackBot (https://www.flirkorea.com/products/packbot/?vertical=ugs&segment=uis), RS1-TS (http://becauserobots.org/links/2017/05/robot-firefighter/), LUF 60 (https://www.chasesiam.com/products/b5-fire-fighting-vehicle/fire-fighting-robot/luf-60-fire-fighting-robot/), BEAR (https://www.army.mil/article/48456/robots_to_rescue_wounded_on_battlefield), Hibiscus (https://furo.org/en/works/hibiscus.html) 69
Fig. 5.3 Domestic disaster response robots (*Source* https://www.inews24.com/view/115129) by DT3, Securo (https://www.clien.net/service/board/news/277399), LUF60 (https://www.incheon.go.kr/119/NE030401/2072520), Fire Control robot (https://www.hyundai.co.kr/news/CONT0000000000163672), RobHaz (https://m.post.naver.com/viewer/postView.naver?volumeNo=39911289&memberNo=44743697), ARMstrong (https://www.kaeri.re.kr/board/view?pageNum=1&rowCnt=10&no1=1301&linkId=11617&menuId=MENU00326&schType=0&schText=&boardStyle=Image&categoryId=&continent=&country=&schYear) 70
Fig. 5.4 Cleaning robot platforms (*Source* https://www.directindustry.com/ja/prod/adlatus-robotics-gmbh/product-200716-2023748.html) by Adlatus CR700, VIGGO S100-N (https://korean.alibaba.com/product-detail/S100N-Autonomous-Robot-Sweeper-Cleaner-Intelligent-1601200323636.html) 71
Fig. 5.5 Cleaning robots LEOBOTS (*Source* https://www.lionsbot.com/leobots/) 71
Fig. 5.6 Autonomous wheelchair 73
Fig. 5.7 Autonomous wheelchairs and advanced navigation systems for individuals with limited mobility 73
Fig. 5.8 Following robot design process 74
Fig. 5.9 Following robot that drives outdoors 74
Fig. 5.10 Autonomous delivery robot (*Source* https://www.robotis.com/sub/business_012.php) by Robotis, Woowa Brothers (https://robot.baemin.com/) 76

Fig. 5.11 Autonomous tractor (*Source* Qu et al., 2024) 77
Fig. 5.12 Automated fruit harvesting robot (*Source* Onishi et al., 2019) 77
Fig. 5.13 Bricklaying robot SAM100 (*Source* Dakhli & Lafhaj, 2017) 78
Fig. 5.14 Industrial robots (*Source* Dzedzickis et al., 2021) 78
Fig. 5.15 Pedestrian environment assessment using autonomous robots 79
Fig. 5.16 Geo-MMS UAV with Lidar units (*Source* Jing et al., 2022) 80
Fig. 5.17 Lidar point cloud segments captured for railway mapping (*Source* Jing et al., 2022) 80
Fig. 6.1 Robot services in smart cities. Security robots, as shown in **a**, are deployed in public spaces to patrol and monitor for potential security threats. This image is sourced from Knightscope (https://www.knightscope.com/use-cases/law-enforcement). Delivery robots, depicted in **b**, are used for logistics and food delivery within smart cities, enhancing efficiency in urban environments. This image is sourced from Starship Technologies (https://www.starship.xyz/). Medical robots, represented in **c**, such as LG CLOi, perform tasks like visitor guidance, medical supply delivery, and nighttime patrols within hospitals, offering personalized services. This image is sourced from LG Electronics (https://live.lge.co.kr/lg-cloi-2/). Guide robots, shown in **d**, assist users in navigating complex spaces such as shopping malls and airports, improving accessibility and convenience. This image is sourced from LG Electronics (https://live.lge.co.kr/lg-cloi-guidebot/) 85
Fig. 6.2 Autonomous Delivery Robot (ADR) routing with public transportation. Schematic of ADRs integrated with public transportation for smart city logistics, enhancing delivery range, efficiency, and sustainability (De Maio et al., 2024) 89
Fig. 6.3 Mobility hub flow 89
Fig. 6.4 Concept of mobility hub 90
Fig. 6.5 Integration of autonomous robots in mobility hubs 91
Fig. 6.6 Autonomous valet parking robot system 91
Fig. 6.7 Changes in urban spatial structure with the introduction of autonomous robot services (https://www.hyundai.com/worldwide/en/newsroom/detail/hyundai-motor-honored-with-inaugural-autoweek-vanguard-award-0000000324) 94

Fig. 6.8 Autonomous Robots in the Logistics Industry. This figure highlights the application of autonomous robots transforming the logistics industry. Panel **a**, sourced from Locus Robotics (https://locusrobotics.com/), shows mobile robots optimizing warehouse operations by efficiently navigating and transporting items. Panel **b**, sourced from Grey Orange (https://www.greyorange.com/), illustrates an automated storage and retrieval system (ASRS), where robots enhance storage efficiency and accelerate item retrieval, demonstrating cutting-edge goods-to-person automation 95
Fig. 6.9 Vision for the Gwangmyeong-Siheung mobility hub city 96
Fig. 7.1 Ethical decision of autonomous robot 101
Fig. 7.2 Image collection of autonomous driving car 102
Fig. 7.3 Autonomous driving delivery (Srinivas et al., 2022) 108
Fig. 7.4 PM playground service complex design (Choi et al., 2023) 109
Fig. 7.5 Social consensus for social acceptance 111
Fig. 8.1 Robots driving sustainability: advancing 13 of 17 UN SDGs through innovation and green technology. *Source* https://ifr.org/ifr-press-releases/news/robots-help-reaching-un-sdgs 114
Fig. 8.2 SpaceX's Falcon9 rocket with reusable booster. *Source* https://www.seattletimes.com/business/faa-grounds-spacex-after-rocket-falls-over-in-flames-at-landing/ 116
Fig. 8.3 Circular economy: end-of-life products are recycled and recovered instead of becoming residual waste. *Source* https://www.triumvirate.com/blog/do-you-know-about-the-circular-economy 118
Fig. 8.4 A framework for robot design considering lifespan and sustainability 119
Fig. 8.5 Autonomous robot with solar panels. *Source* https://www.yankodesign.com/2021/08/17/this-medicine-delivery-bot-carries-your-supplies-in-the-last-mile-using-a-solar-powered-drone/ 120
Fig. 8.6 Framework for life cycle assessment (from ISO 14040:2006; modified). *Source* Wolf et al. (2010) 122
Fig. 8.7 Global electricity demand from data centres, AI, and cryptocurrencies, 2019–2026. *Source* Electricity (2024) 123
Fig. 8.8 Estimated electricity demand from traditional data centres, dedicated AI data centres and cryptocurrencies, 2022 and 2026, base case. *Source* Electricity (2024) 123
Fig. 8.9 Edge-driving robotics platform (EDRP) and local dynamic map platform (LDMP). *Source* Moon et al. (2024) 124

Fig. 8.10 Typical data center energy usage. *Source* ABB Motors and Mechanical Inc. (2022) 125
Fig. 8.11 Understanding data center dynamics for optimizing PUE. *Source* https://blog.google/inside-google/infrastructure/better-data-centers-through-machine/ 126
Fig. 8.12 Understanding data center dynamics for optimizing PUE. *Source* https://deepmind.google/discover/blog/deepmind-ai-reduces-google-data-centre-cooling-bill-by-40/ 126
Fig. 8.13 SMP robotics—air quality monitoring robot (S2.3 Eco). *Source* https://smprobotics.com/products_autonomous_ugv/air-quality-monitoring-robot/ 128
Fig. 8.14 KOBOT (KOAI)—autonomous marine pollution control robot. *Source* https://koai.co.kr/kobot-remote-control/ 129
Fig. A.1 Autonomous robot-based framework for pedestrian environment assessment 132
Fig. A.2 Pedestrian environment assessment factors 133
Fig. A.3 WeGo ST-mini 135
Fig. A.4 Size of robot 135
Fig. A.5 Scout mini ROS Topic 139
Fig. A.6 Cameras used in the test autonomous robot 140
Fig. A.7 LiDAR used in the test autonomous robot 141
Fig. A.8 Camera parameter adjustment—rqt_reconfigure_param 142
Fig. A.9 Camera and ground projection image after camera parameter adjustment (simulation) 143
Fig. A.10 Camera and ground projection image after camera parameter adjustment (simulation) 143
Fig. A.11 ROS environment—autonomous driving simulation at University of Seoul 144
Fig. A.12 Sidewalk test section at University of Seoul 145
Fig. A.13 Sidewalk 1 at the University of Seoul 145
Fig. A.14 Sidewalk 2 at the University of Seoul 146
Fig. A.15 Sidewalk 3 at the University of Seoul 147
Fig. A.16 System implementation code for assessing robot damage and obstacles 148
Fig. A.17 Rosbag topic nodes 149
Fig. A.18 RTK-GPS data 150
Fig. A.19 GPS data extracted from ROS (Sidewalk 1) 151
Fig. A.20 Post-processing of GPS coordinates after autonomous driving command 151
Fig. A.21 Obstacle detection (robot's point of view) 152
Fig. A.22 Surrounding obstacle detection 152
Fig. A.23 Robot driving 152
Fig. A.24 Robot obstacle avoidance 153
Fig. A.25 Obstacle detection on Sidewalk 1 (robot's point of view) (1) 154
Fig. A.26 Obstacle detection on Sidewalk 1 (robot's point of view) (2) 155

Fig. A.27 Obstacle (PM) avoidance and autonomous driving (1) 156
Fig. A.28 Obstacle (PM) avoidance and autonomous driving (2) 157
Fig. A.29 GPS data extracted from ROS (Sidewalk 2) 158
Fig. A.30 Post-processing of GPS coordinates after autonomous driving command 158
Fig. A.31 Surrounding obstacle detection 159
Fig. A.32 Obstacle detection 159
Fig. A.33 Robot driving 159
Fig. A.34 Robot obstacle (pedestrian) avoidance 160
Fig. A.35 Recognition of learned obstacles on Sidewalk 2 (robot's point of view) (1) 161
Fig. A.36 Recognition of learned obstacles on Sidewalk 2 (robot's point of view) (2) 162
Fig. A.37 Obstacle (pedestrian) avoidance and autonomous driving (1) 163
Fig. A.38 Obstacle (pedestrian) avoidance and autonomous driving (2) 164
Fig. A.39 GPS data extracted from ROS (Sidewalk 3) 165
Fig. A.40 Post-processing of GPS coordinates after autonomous driving command 165
Fig. A.41 Obstacle detection (PM) 166
Fig. A.42 Obstacle detection (bicycle) 166
Fig. A.43 PM avoidance 167
Fig. A.44 Bicycle avoidance 167
Fig. A.45 Recognition of learned obstacles on Sidewalk 3 (robot's point of view) (1) 168
Fig. A.46 Recognition of learned obstacles on Sidewalk 3 (robot's point of view) (2) 169
Fig. A.47 Sequence of obstacle (PM, bicycle) avoidance and autonomous driving (1) 170
Fig. A.48 Sequence of obstacle (PM, bicycle) avoidance and autonomous driving (2) 171
Fig. A.49 Sequence of obstacle (PM, bicycle) avoidance and autonomous driving (3) 172
Fig. A.50 Obstacle classification results using the robot 173
Fig. A.51 Error example—actual: person, predicted: PM (top: detection result, bottom: original image) 174
Fig. A.52 Error example—actual: car, predicted: person (top: detection result, bottom: original image) 175

Fig. A.53 Error example—actual: light, predicted: tree (top: detection result, bottom: original image) 176
Fig. A.54 Surface damage classification results using the robot 177
Fig. A.55 Accuracy and variation in damage detection according to directional change 177

List of Tables

Table 1.1 Existing competitive or similar books 3
Table 1.2 SAE's autonomous driving level 4
Table 1.3 What companies are researching and developing in autonomous driving 5
Table 1.4 Autonomous robotics applications and leading companies ... 15
Table 1.5 Trends in pedestrian walkway autonomous robotics utilization and technology 21
Table 8.1 Mitigation strategies for data centers 127
Table A.1 Specifications and features of the RealSense camera 135
Table A.2 Specifications and features of the RealSense camera 136
Table A.3 Specifications and features of VLP-16 137
Table A.4 Specifications and features of MRP 2000 137
Table A.5 Specifications and features of VN100 138
Table A.6 ROS topic 139
Table A.7 Roles of cameras mounted on autonomous robots 139
Table A.8 Roles of LiDAR mounted on autonomous robots 140
Table A.9 Characteristics of Sidewalk 1 at the University of Seoul 145
Table A.10 Characteristics of Sidewalk 2 at the University of Seoul 146
Table A.11 Characteristics of Sidewalk 2 at the University of Seoul 146

Chapter 1
Systematic Analysis of the Need for Autonomous Robots

1.1 Introduction

Advances in transportation have fundamentally changed the way humans move around. It has revolutionized the economic, social, and cultural spheres beyond just transportation. Initially, transportation began as a luxury item, but through mass production, it became a universal means of transportation. This facilitated logistics, person-to-person travel, and interaction between nations, leading to the rapid growth of industry and economies around the world.

In the 2000s, technological advances accelerated even further. In particular, rapid advances in computer science and artificial intelligence have opened up new possibilities for the transportation industry. The mechanical functions inside the vehicle became increasingly intelligent and began to be equipped with systems that incorporated artificial intelligence. This transformation has led to the evolution of the vehicle from a simple transportation tool to an autonomous vehicle that maximizes safety and convenience for the driver, and eventually to self-driving technology that makes decisions and moves without a driver. In recent years, autonomous technology has joined forces with the transportation industry to create growth and development that was never possible before. Autonomous vehicles are now moving beyond the experimental stage and are being tested on real-world roads, leading to a shift in policy away from road-centric policies and toward environmental protection, public transportation, and pedestrian-centric policies. These policy shifts are having a positive impact on urban planning and transportation systems, contributing to improved pedestrian safety and urban sustainability. These changes are illustrated in Fig. 1.1.

Key elements of autonomous driving technology include advanced artificial intelligence, computer vision, and sensor technology. These technologies enable vehicles or robots to accurately perceive their surroundings and navigate safely in complex road conditions. Additionally, V2X technologies, which connect everything, including vehicle-to-vehicle (V2V) and vehicle-to-infrastructure (V2I) communications, are helping vehicles to operate smarter. The integration of these technologies

M. Choi and S. Lee, *Intelligent Moving Cities: Technological Leap and Social Integration of Autonomous Mobile Robots*,
https://doi.org/10.1007/978-981-96-8864-7_1

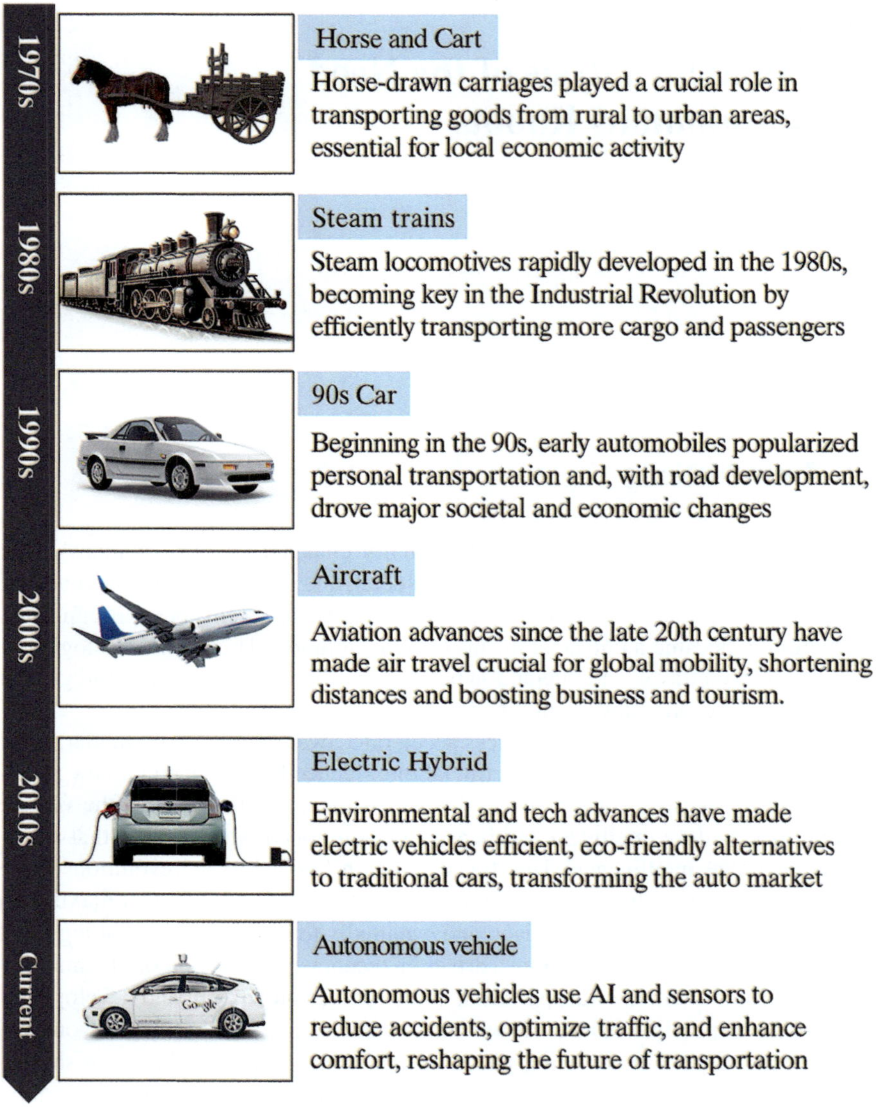

Fig. 1.1 Changes in transportation

means that autonomous driving is not just a meaningful development within the transportation industry, but a major pillar in the transformation of our entire social fabric and way of life.

Existing books on autonomous robots focus primarily on technical aspects and the potential integration of robots into urban environments. They discuss the fundamental impact that autonomous robots can have on society and urban infrastructure, including their mobility, sensing, and navigation technologies. They also explore the ethical and social aspects of autonomous robots and outline how technological advances can be integrated into human-centered urban environments. A comparison of these existing books is provided in Table 1.1.

While previous books have outlined the integration of autonomous vehicles into urban environments, this book explores the role and possibilities of autonomous technology at the intersection of multiple technological advances and social change. We look at how this technology can transform our daily lives and provide solutions to social, economic, and environmental challenges. The development of autonomous driving technology is more than just a technological achievement; it is becoming an important vehicle for positive change in our society and environment.

Table 1.1 Existing competitive or similar books

Book	Author(s)	Year	Comparison with this book
Autonomous mobile robots and multi-robot systems: motion-planning, communication and swarming	Eugene Kagan et al.	2019	Provides a more holistic approach, considering the broader impacts of autonomous mobile robots on urban environments and societies
Introduction to autonomous robots	Mattias Wahde	2016	Takes a broader approach to the impact of autonomous robots on the urban environment and society, specifically exploring how they can contribute to urban planning and pedestrian-oriented development
Introduction to autonomous mobile robots	Roland Siegwart et al.	2011	Has a different approach and objectives for autonomous mobile robots, with a greater focus on the integration of autonomous mobile robotics technology within the urban environment and its social inclusion
Essential principles for autonomous robotics	Henry Hexmoor	2023	Provides more in-depth analysis and discussion of how autonomous robots can promote human-centered development within the urban environment
Autonomous mobile robots: sensing, control, decision making and applications	Frank L. Lewis et al.	2006	Greater focus on the integration of autonomous mobile robots in urban environments and their societal impacts
Autonomous mobile robots: planning, navigation and simulation	Rahul Kala	2023	About the use and advancement of autonomous robots, but with deeper technical content related to the execution of the technology
Mobile robotics	Luc Jaulin	2019	Describe the design considerations for each of the technologies used in autonomous robots and derive state equations through robot modeling

1.2 The Intersection of Technological Progress and Social Demand

Autonomous driving technology is the technology that allows robots or cars to drive themselves without driver intervention. To eliminate driver intervention, the vehicle body must have functions that replace the driver's driving behavior, such as collision avoidance, emergency braking, cruise control, and lane departure warning. It also requires improved sensors for object recognition, high-performance graphics processing units, odometers, and GPS to replace the driver's thought processes that recognize the environment around them. The level of development of autonomous vehicles is typically categorized into six stages to align with the advancements and stages of these technologies. The Society of automotive Engineers (SAE) has divided them into stages 0 through 5, as summarized in Table 1.2 and 1.3 and illustrated in Fig. 1.2.

Research into self-driving cars is not just limited to the automotive industry (Hyundai, Toyota, Mercedes-Benz, BMW, Ford, etc.), but also global tech companies (Google, Nvidia, etc.).

The history of self-driving cars begins in the 1920s. Francis Houdina, an electrician in the US Army, developed the American Wonder, a car that could be controlled wirelessly via radio signals. Later, in the 1930s, self-driving cars were created by laying RF modules on the road or drawing specific patterns and following them. The car that embodies the current concept of autonomous driving appeared in Japan in 1977.

Table 1.2 SAE's autonomous driving level

Level	Name	Definition	Featured technologies
Level 0	No automation	• All operations are performed by the driver	• No autonomous driving
Level 1	Driver automation	• Driver: driving system: perform specific driving modes of steering, acceleration, and deceleration	• Maintaining following distance Lane Keeping
Level 2	Partial automation	• System: driving driver: responds directly to system requests for intervention, always ready to take control of the vehicle	• Following distance + lane keeping autopark
Level 3	Conditional automation	• Systems: Driving Allows the system to drive even if the driver fails to respond appropriately to intervention requests	• Autonomous driving on highways
Level 4	High Automation	• Systems: driving • Allows the system to drive even if the driver fails to respond appropriately to intervention requests	• For specific segmentsand weather conditions
Level 5	Full automation	• System always on in all conditions	• Autonomous driving • Driverless transportation

Table 1.3 What companies are researching and developing in autonomous driving

Company	Research and develop
Hyundai Motor Company	Level 3 highway driverless piloting technology (HDP). Development of a Level 4 robotaxi based on IONIQ5
Toyota Motor Corporation	Registered the highest number of autonomous driving technology patents
Mercedes-Benz AG	World's first international certification of Level 3 conditional autonomous driving system 'Drive Pilot'. Level 4 automated parking technology
BMW	Leverage cameras, radar, and ultrasonic sensors to continuously monitor the vehicle's surroundings
Ford Motor Company	Active in developing self-driving systems, including the acquisition of artificial intelligence startup Argo AI
Google	The world's first self-driving car is in the works (Waymo)
Nvidia	Build an autonomous vehicle development platform, including hardware and software

These cars followed specific markings. Since the 1980s, autonomous driving systems have become more sophisticated, starting with the development of camera sensors, and today, advances in computer technology, artificial intelligence, and communication technology have led to the development of more robust autonomous driving systems.

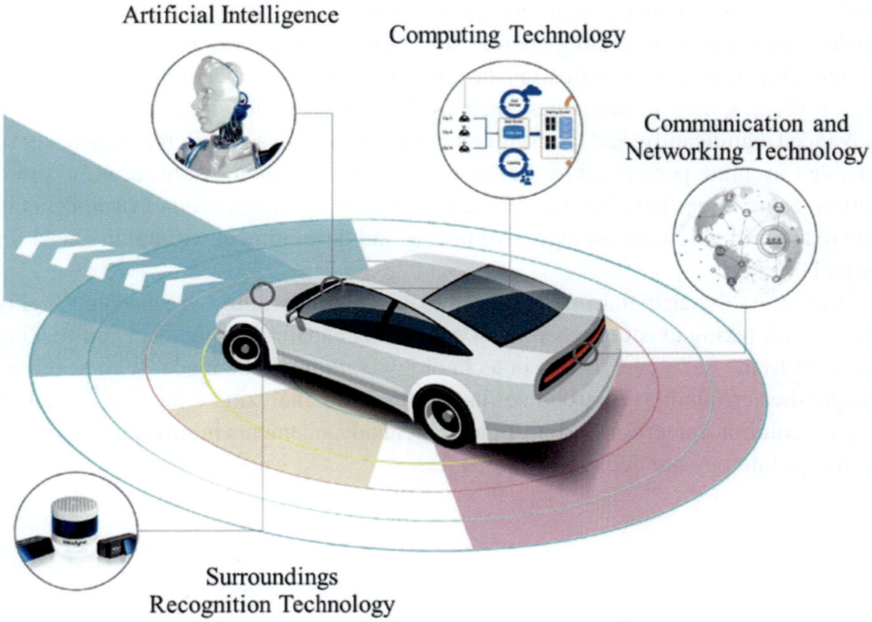

Fig. 1.2 Technologies in autonomous robots

1.2.1 Surroundings Recognition Technology

Since before the twenty-first century, the research community has been racing to develop safe autonomous vehicles capable of real-world high-speed driving. According to Van Brummelen et al. (2018), one of the first road tests of autonomous vehicles, 'No Hands Across America', was introduced in 1995 (Bertozzi et al., 2000). It is notable for testing the performance of autonomous vehicles across the United States. Around the same time, the Munich to Odense UBM Test is known to have tested autonomous driving from Germany to Denmark. In 1998, the ARGO project completed testing in Italy in a variety of climates, validating lane departure warning systems, lane keeping systems, and speed control systems. Through these tests, researchers were able to identify areas for improvement in autonomous driving technology. Areas of improvement include image processing and recognition technology, driving in adverse weather conditions, and improved obstacle and road sign detection.

The 2007 DARPA Urban Challenge tested several urban driving situations, including intersections, highway on-ramps, and off-road. To represent real-world urban driving conditions, the autonomous vehicles obeyed traffic laws (Rouff, 2012). However, they were not tested for their response to moving obstacles such as pedestrians or bicycles. In addition, there were problems with detecting traffic lights and signs. To address these issues, the VisLab Intercontinental Autonomous Challenge was launched in 2010.

It tested the real-time performance of autonomous vehicles on off-road, highway, and in various weather conditions. Real-world driving has greatly increased our understanding of surroundings perception, obstacle detection and avoidance, and provided critical data for evaluating the reliability and safety of autonomous systems. Since then, sensor technologies such as lidar, radar, and camera systems have advanced significantly, and the complementary capabilities of different sensors have enabled accurate perception of the surrounding environment. More recently, simulation technologies have been developed to simulate complex road environments in the real world, allowing for extensive testing and development without the need for actual driving.

In the 2020s, artificial intelligence and machine learning have become highly developed, taking existing sensor data processing capabilities and decision-making systems to the next level. Leaps in technology continue to be made, such as recognizing the surrounding environment in real time and analyzing images collected by high-resolution cameras in a short time. Recent advancements in surrounding recognition technology are depicted in Fig. 1.3.

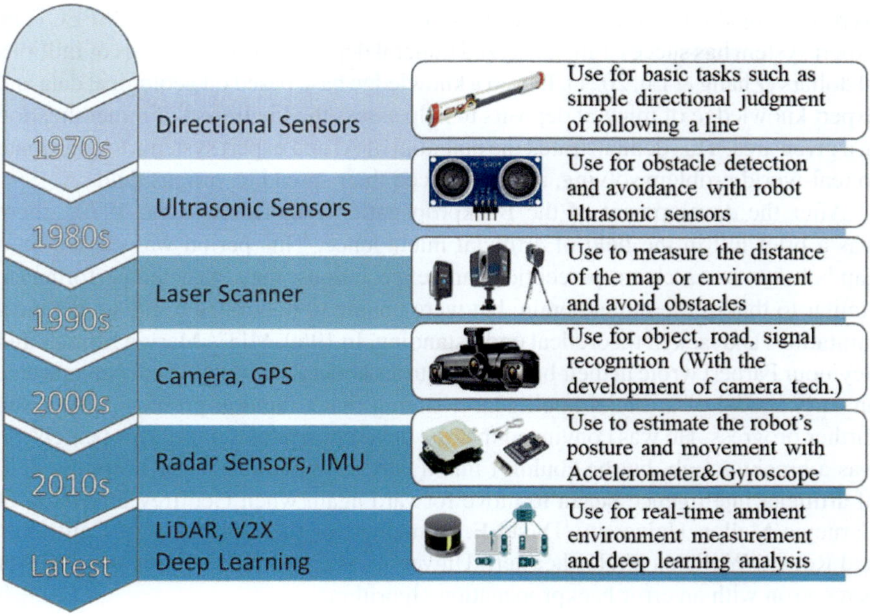

Fig. 1.3 Advances in surrounding recognition technology

1.2.2 Artificial Intelligence

As mentioned earlier, the development of artificial intelligence is highly correlated with the development of autonomous driving technology. It's used for perception and information processing, but it's also used for decision-making and vehicle-to-everything (V2X) communication technologies. The complex decision-making process requires the ability to make optimal decisions under a variety of conditions, which is achieved through artificial intelligence. In addition, the information gathered must be analyzed quickly while communicating with other vehicles and city infrastructure, so having an efficient artificial intelligence system is essential for autonomous driving. In addition to communicating with its surroundings and recognizing the environment in real time, artificial intelligence plays a huge role in the advancement of autonomous driving. It can predict traffic conditions and set safe and efficient driving routes. This is crucial for the overall transportation system when there are multiple autonomous vehicles in a city. In addition, simulation is essential in the development and validation of autonomous driving technology, and AI can help ensure that the simulations reflect the appropriate driving algorithms and allow for a wide range of situations.

Artificial intelligence is the study of how to get computers to perform intelligent tasks that only humans have been able to perform in the past (Huang et al., 2019). AI has driven economic development and productivity. It evaluates large amounts of data and helps in decision-making. Expert systems, an early model of AI in the 1970s, were also popular for simplifying decision-making. For example, the PROSPECTOR expert system has successfully analyzed mineral deposits worth hundreds of millions of dollars (Zhang & Lu, 2021). It used a knowledge base based on geological data and expert knowledge of mineral deposits to help assess the likelihood of mineralization in a given area. This demonstrated the potential of AI and expert systems to contribute to real-world problem solving, and influenced their spread to various fields.

After the development of the Backpropagation Algorithm in the 1970s, there was a brief lull in the field of artificial intelligence. This period was caused by a gap between researchers' perceptions and expectations: they expected performance similar to that of the human brain, but were unable to maximize it due to technical limitations and lack of theoretical understanding. In 1969, MIT's Marvin Minsky and Seymour Parpert wrote in their book Perceptrons about the limitations of perceptrons, an early model of artificial intelligence, stating that structural problems prevented further progress. He was convinced that the development of a multilayer perceptron was a breakthrough, but he couldn't find a way to implement it. In 1986, the field of artificial intelligence began to move forward again when Geoffrey E. Hinton of Carnegie Mellon University, David E. Rumelhart of the University of California, and Ronald Williams of Northeastern University succeeded in training a multilayer perceptron with an error backpropagation algorithm.

In 1980, Kunihiko Fukushima of the NHK Broadcasting Technology Research Institute developed a neural network model called the "Neocognitron". It was based on medical findings from the 1950s and structured how information received by the eye is processed by neurons in the brain. In 1989, Yann LeCun of AT&T Labs developed a program to recognize and automatically sort handwritten postal codes. It was able to recognize handwriting with about 90% accuracy. LeCun and his colleagues developed the Neocognitron by applying an error backpropagation algorithm, and in 1998, they published the Convolutional Neural Network (CNN). In addition, the success of Deep Blue, a chess-playing computer developed by IBM in 1997, catapulted AI development into the spotlight. The increase in computing power accelerated the pace of AI development, and the development of big data-based deep learning and reinforcement learning began in earnest. However, despite the successful development of CNN, it had to face another downturn due to the slow learning speed and inaccurate learning as the neural network became more sophisticated.

This was improved in the late 2000s. The accuracy problem of learning was solved by using the ReLU function as the neuronal activity function. The advantage over the traditional sigmoid function was that intermediate computational results could be transferred directly. The slower learning speed was offset by the dramatic improvement in computer hardware performance, such as GPUs.

In the 2010s, the field of deep learning was revitalized by the availability of large datasets and superior hardware. Various neural network architectures such as RNN, LSTM, and others have been developed in addition to CNN, and they have shown effective performance for specific problems ranging from image recognition to natural language processing and time series data analysis. The field of artificial intelligence, especially deep learning, has grown rapidly as solutions to problems such as gradient loss and overfitting in deep learning have been researched. Recently, it has been used for face recognition, object detection and recognition in images, etc.

1.2.3 Computing Technology

Along with artificial intelligence, advances in computing technology have played a large role in the recent rapid expansion of the autonomous driving industry. This makes sense, as it requires hardware assets that can handle machine learning and deep learning. Computing technology advancements are highly correlated with AI, sensor data processing, and simulation sophistication. Today's CPU and GPU performance, the culmination of nearly 100 years of computer advancements, is playing a critical role in the development of autonomous robot hardware and systems. The large amount of data generated by autonomous robots every second must be stored and managed for learning, analysis, and system validation, and data centers and cloud storage are also key areas for this.

The Turing machine, the predecessor of the computer, was invented by Alan Turing in 1937. It is an abstract computational model that allows any computational process to be represented mechanically, which later provided the theoretical basis for digital computers. It influenced the logical structure of programmable computers.

John Atanasoff, a physics professor at Iowa State University, and his student Clifford Berry developed the Atanasoff-Berry Computer (ABC) in 1942. It was the world's first calculator to use vacuum tubes and demonstrated the potential of binary-based computation and electronic data processing. It was designed specifically for solving linear algebra problems. Although the process of using vacuum tubes to process information became a key principle in later developments of electronic computers, it had the disadvantage that programs could not be embedded and executed, and the machine had to be set up manually. Nevertheless, the ABC project is recognized as an important milestone in the early history of computer science. Two years later, in 1944, one of the world's first large-scale automatic digital computers, the MARK-I, was developed at Harvard University in collaboration with Howard Aiken and IBM. An electromechanical automatic calculator, it was huge, measuring 15.5 m long, 2.4 m high, and 0.6 m wide. It had the ability to perform complex calculations, including logarithmic, trigonometric, and exponential calculations, and was used primarily for military purposes, particularly for complex calculations such as bullet trajectory calculations for the Navy. The development and successful operation of the MARK-I is considered an important step in the development of the electronic computer and was a cornerstone of programming language development (Fig. 1.4).

Completed in 1946, ENIAC was the world's first general-purpose electronic digital computer, developed by John Mockley and J. Presser Eckert, Jr. at the University of Pennsylvania. ENIAC was designed to calculate the trajectory of artillery shells during World War II and used 18,000 vacuum tubes. Developed in 1949 at the University of Cambridge, England, EDSAC was the world's first stored-program digital

Fig. 1.4 Mark I, the world's first large-scale digital computer (Getty Images by PhotoQuest)

computer designed for real-world tasks. It was used for research and scientific calculations and is considered one of the precursors to the modern computer because of its ability to store programs in memory. The UNIVAC-I, released in 1951, was the world's first commercially successful digital computer, designed for the U.S. Census Bureau. Developed by Jean Ordinary and Presser Eckert, the machine was used for a variety of private and governmental purposes, including election forecasting. The success of the UNIVAC I proved the commercial viability of the computer industry.

The Electronic Discrete Variable Automatic Computer (EDVAC), completed in 1952, was the world's first computer to use a stored program approach. It was a successor to the earlier ENIAC, developed as part of a project led by John Mockley and J. Presser Eckert, Jr. at the University of Pennsylvania. Von Neumann laid out the basic principles of computer design for how computers store programs in memory, known as the "von Neumann architecture," which is still used in most computers today. The EDVAC used binary numbers to perform calculations and could store both instructions and data in memory. This allowed various operations to be performed through software without the need to physically change the program. This was a big improvement over the ENIAC's manual setup with plugs and switches.

The IBM System/360 is a computer system introduced by IBM in 1964. It is considered to be one of the first computer architectures with compatible software and hardware. The System/360 was available in models of various sizes and capabilities to meet the needs of a wide range of customers, from small businesses to large research organizations. One of the biggest innovations of the system was that all models shared a single architecture and instruction set. This made software and applications compatible between the different models, making it much simpler to upgrade or replace computer systems as businesses grew or their needs changed. The System/360 had the versatility to be used for both business data processing and scientific computations, which was a major differentiator compared to other systems of the time. The system set the standard for mainframe computers and influenced many areas of computer science, including software development, system administration, and database management. The introduction of the IBM System/360 was a huge success in the computer industry, and the system was used by companies and organizations around the world for decades. The system is considered a major milestone in the development of modern computing, and its legacy continues today in IBM's z-series mainframe computers.

In 1971, the Intel 4004 was released as the world's first microprocessor. It was an essential technological advance in the evolution of the personal computer and revolutionized the world of computing. In the mid-1970s, companies like Apple, Altair, and Microsoft began developing personal computers or software languages. The Apple I and Apple II were a huge success in the market because they allowed users to assemble them at home and introduced an easy-to-use GUI.

The personal computer market really began to expand when IBM introduced the PC in 1981. Many manufacturers began releasing computers that were compatible with the IBM PC, and Apple's Macintosh introduced a GUI and mouse, creating a more user-friendly computer environment. As the PC evolved, office software appeared, and the concept of the Internet began to expand. Along with the PC, research into computer vision, sensor technology, and more began, providing an important technological foundation for the development of modern autonomous vehicles and robots.

In the 1990s, GPS technology was commercialized to provide accurate location information, and advances in wireless communication technology expanded the utility of computers. In particular, the speed at which data can be transmitted and received began to dramatically change, playing an important role in robot collaboration and remote control.

In the 2000s, research on autonomous robots progressed as sensor technology and artificial intelligence learning algorithms were researched alongside advances in computer technology. This was the time when autonomous vehicle racing competitions were held and full-scale commercialization research began. Since the 2010s, the field has expanded to include the commercialization of artificial intelligence, machine learning, and autonomous driving technology, and advances in high-performance computing hardware such as GPUs have enabled explosive performance improvements. There has been dramatic growth in CPUs, GPUs, and cloud computing, making it possible to store and process large amounts of data. As computers became capable of computing more complex algorithms and shorter processing times, other cutting-edge fields were able to make leaps and bounds.

1.2.4 Communication and Network Technology

Communication technology has a history of continuous innovation, from the invention of the automobile to the present day. Starting with early analogue communication

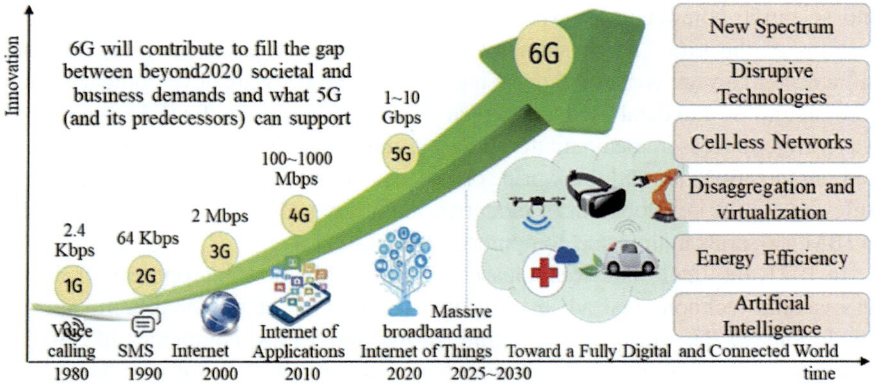

Fig. 1.5 Evolution of telecom networks from 1 to 6G

methods, it evolved with the advent of the digital age to technologies that can transmit complex data faster, farther, and more reliably. The evolution of cellular communication technology from 1 to 5G has exponentially increased the speed of information transfer, which has fundamentally changed autonomous driving as shown in Fig. 1.5.

Autonomous roads need to collect real-time information about their surroundings, make judgments and decisions based on it. Communication technologies such as vehicle-to-vehicle (V2V), vehicle-to-infrastructure (V2I), and vehicle-to-everything (V2X) play a key role in this process (Khan, 2021). These communication technologies enable vehicles to receive data from surrounding vehicles and road infrastructure, analyze it, and use it to prevent collisions, optimize traffic flow, and provide information to drivers.

The ARPANET, a packet-switched network created by the U.S. Department of Defense in the 1970s, was an early model of the Internet that began as a military project. The idea was to install servers in different locations and connect them to each other. Once the network was connected in the U.S., universities began experimenting and sharing data for academic research. As more universities joined, the network expanded, connecting the world and evolving into the current Internet.

In the 1980s, the world's first cellular networks were built, enabling mobile telephony, and TCP/IP, the core protocol of the Internet, was introduced. In the 1990s, conditions for the Internet and mobile communications improved significantly, accelerating the development of the Internet and mobile communications, including the launch of the World Wide Web, the commercialization of 2G mobile communications, and the spread of fiber optic technology. From 2000 to the present, the development of 3G to 5G mobile communications and the development of Wi-Fi have provided the background for individual network use. With the recent development of 6G, the development of SW-centered autonomous vehicles is becoming more active.

The commercialization of 5G technology and the development of 6G technology are strongly linked to the advancement of autonomous driving. In order to achieve full automation, where the system is in charge of all procedures and driving, massive amounts of data must be exchanged without delay. In order to achieve full automation, where the system is in charge of all procedures and driving, a large amount of data must be exchanged without delay. This is why 6G technology, which can transmit large amounts of data quickly, is essential. It can also be utilized for unmanned vehicles such as unmanned ships, UAMs, and drones. It is not only the communication devices or performance of the car that is important, but also the communication performance of the infrastructure around the road. In particular, it is much more stable in terms of computer efficiency and power consumption if the surrounding data collected by existing state-of-the-art sensors is delivered through a communication network.

Autonomous vehicle communication includes direct vehicle-to-vehicle (V2V) communication, vehicle-to-infrastructure (V2I) communication, vehicle-to-pedestrian (V2P) communication, and, more recently, vehicle-to-x (V2X), a concept that connects the vehicle to everything around it. By utilizing communication technology, drivers can recognize much more information than they can get by looking at their surroundings from inside the car, and it can help prevent accidents. However, one of the obstacles to the development of V2X technology has been the slow adoption of international standardization. Without a standardized method, the direction of technology development is confusing and the process of commercialization and verification is slow.

Advances in global positioning system (GPS) technology have played an important role in the development of autonomous driving technology. Early GPS systems had a positioning accuracy of about 10 m, but high-precision GPS can pinpoint locations to a meter or less. These GPS signals are affected by various factors in the atmosphere, and the technology to compensate for this has been developed and improved. Recently, satellite technology for fast communication has been pioneered by the automaker Tesla. With Starlink, Tesla is building a new paradigm of satellite systems that will enable internet network connectivity in areas without cell towers. With Starlink, a low-orbit satellite, Tesla aims to enable uninterrupted satellite communications in areas with low cell tower density, or in areas that are temporarily underserved by natural disasters or war. The system will enable internet speeds that are much faster than the typical 4G and 5G communication speeds we currently use, and is expected to enable seamless internet access even in the middle of the ocean.

1.2.5 Autonomous Driving Applications

Autonomous driving technology is operated by software and hardware rather than a human driver (Deb et al., 2018). Therefore, the social acceptability of autonomous robots is strongly related to the reduction of accident rates and the number of accidents. In order to reduce pedestrian risk exposure, multidisciplinary transportation research and advancement of various functions of autonomous vehicles are needed. Pedestrian Protection System (PPS) can detect the presence, location, speed, etc. of other robots or objects and can effectively control driving to reduce pedestrian accidents (Zhang, 2016). In addition to the ability to recognize and slow down pedestrians, functions that allow pedestrians and vehicles to interact are also being developed. Some researchers are developing methods to detect pedestrians' faces, gaze, and hand gestures, and others are developing manuals to raise their hands when crossing the street in non-crosswalks (Mirnig, 2017).

According to Pettigrew (2018), autonomous robots are also expected to play an important role in terms of health. They are expected to reduce accidents, reduce emissions, improve mobility for people with driving disabilities, reduce stress, and increase cycling. These effects will be maximized when all vehicles are autonomous. However, surveys have shown that the health benefits are not clearly perceived by citizens. Communicating the reduced accident rates, increased safety, and economic estimates of autonomous vehicles will increase public acceptance.

Robotic systems are also being utilized in the agricultural sector to help increase productivity and efficiency in farming (Rahmadian & Widyartono, 2020). Expectations include increased quality and yield, and reduced unharvested waste. Agricultural robots combine the latest technologies in communications, sensor systems, navigation, and geographic information. On farms, autonomous robots apply pesticides or fertilizers and harvest crops. They take into account conditions such as the varying contours, slope, and humidity of the land, allowing for considerable precision farming (Gebbers & Adamchuk, 2010). The ability of a robot to detect its surroundings is a fundamental component in the development of autonomous robots in agriculture, as the robot must move around the field performing multiple agricultural tasks in addition to data collection. In recent years, robotic agricultural systems have become more sophisticated and convenient, thus expanding the applications of autonomous robots (c. Aloisio et al., 2012). Table 1.4 provides details on various applications and leading companies in autonomous robotics.

1.2.6 Demand for Autonomous Driving

Technological advances in the early twenty-first century have revolutionized the way we live and work. The advent of smartphones, in particular, has dramatically improved access to information and changed the way we communicate globally. These changes increased the acceptance of technology across society and paved the way for the development and acceptance of more advanced technologies, such as autonomous mobile robots.

At the same time as smartphones have become increasingly popular, the transition to an aging society has presented new challenges and opportunities for maintaining mobility and independence. Globally, the aging of the population has become an inevitable phenomenon (Chen et al., 2024), which is an important factor driving societal demand for autonomous robots. For people with mobility difficulties, including the elderly, autonomous robots have the potential to help them live independently by

Table 1.4 Autonomous robotics applications and leading companies

Area	Field description	Research from leading companies
Autonomous vehicles	A vehicle that drives itself without driver intervention	• Waymo: Google's self-driving car division, developing innovative self-driving technologies and platforms • Tesla: A leader in electric vehicles and self-driving technology
Logistics and warehouse automation	Move products, manage inventory, or automate tasks within a warehouse	• Amazon Robotics: robotics company for warehouse and logistics automation • Kiva Systems: Developed a logistics automation system that allows robots to move and sort goods within a warehouse
Agricultural automation	Identify and harvest crops	• John Deere: Utilizing GPS and sensor technology to manage crops efficiently • Blue River Technology: Developing autonomous robots and computer vision technology for agriculture to manage and pesticide crops
Construction and architecture	Survey the terrain or transport materials	• Caterpillar: Automating simple tasks with autonomous road and mining machines • Komatsu: Automating mining and construction machinery to perform hazardous tasks on behalf of customers
Medical and health	Monitor patient health or transport medical supplies	• Intuitive Surgical: Develops medical robotic systems to automate surgical processes and advance medical technology • Medtronic: Develops autonomous robotics and surgical robotics technologies used in healthcare
Military and security	Explore hazardous areas or monitor perimeters	• Boston Dynamics: Automating reconnaissance and surveillance tasks • Ava Robotics: Developing robotic systems to patrol and secure the interior of buildings

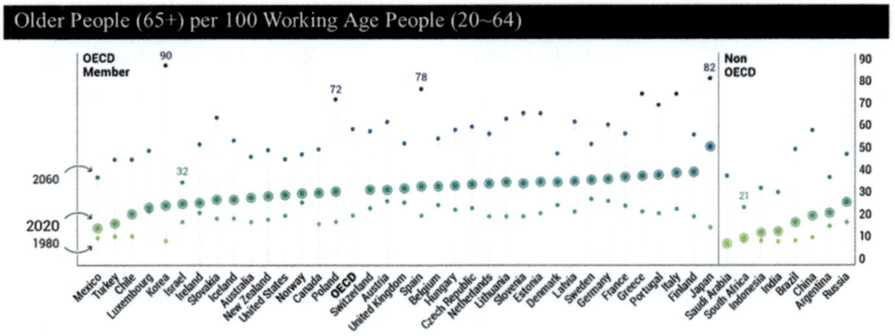

Fig. 1.6 Older population as a percentage of the labor force (*Source* OECD)

providing safe and convenient transportation (Shishehgar et al., 2018). The growing proportion of older adults in the workforce is illustrated in Fig. 1.6.

In addition, the evolution and development of software-defined vehicles (SDVs), the next generation of cars in which software controls and manages the hardware, heralds another shift similar to the socio-cultural change brought about by smartphones. Vehicles are evolving from simple means of transportation to software-controlled, personalized, and constantly updated platforms. This shift will make the acceptance and integration of autonomous robotics easier and reflects a shift in the way we perceive and accept technology. This layered structure of SDVs is depicted in Fig. 1.7.

With the increasing focus on sustainable development, the need for autonomous robots is also becoming more prominent. Autonomous robots have unlimited scope of application (Alatise & Hancke, 2020) and have the potential to revolutionize traffic management and urban planning by optimizing transportation systems, reducing road congestion, and lowering emissions. In addition, they can contribute to reducing negative environmental impacts in the long term. In particular, autonomous delivery robots and patrol robots are practical examples of this change. Autonomous delivery

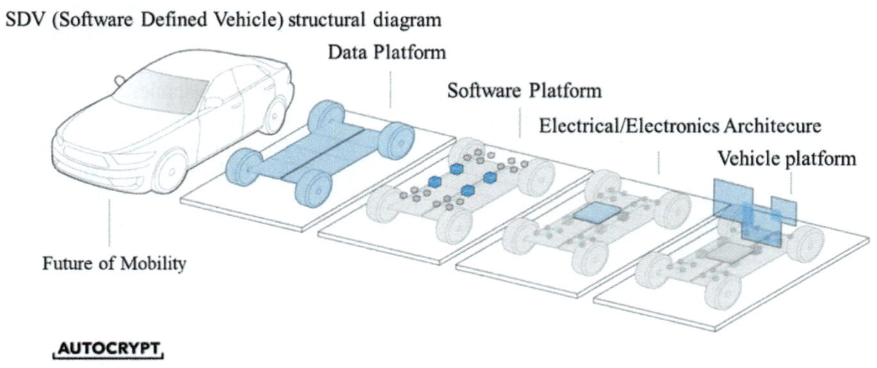

Fig. 1.7 Layers in a software-define vehicles (*Source* AUTOCRYPT)

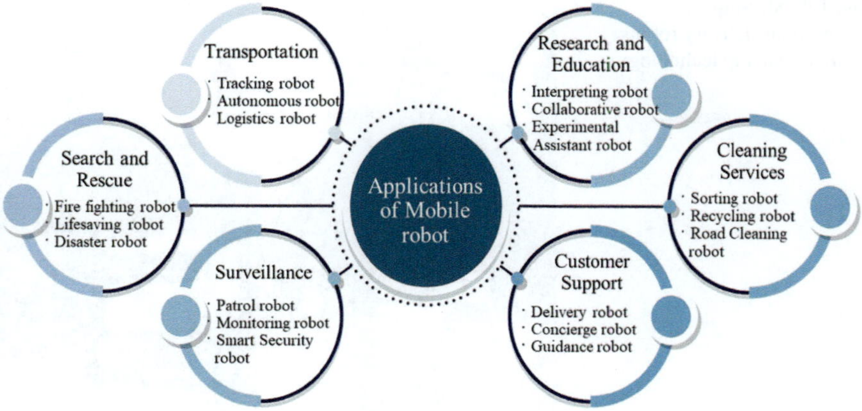

Fig. 1.8 Applications of autonomous robots

robots can reduce traffic on roads by optimizing the delivery of food or goods via sidewalks in urban centers, while patrol robots are expected to play an important role in maintaining public safety with increased security. All of these autonomous robots are expected to offer the benefits of environmentally friendly transportation as summarized in Fig. 1.8.

In addition, recent advances in the technology and legal framework for autonomous robots have significantly expanded the scope of their use from indoor environments to outdoor environments (e.g., sidewalks). This suggests that autonomous technology has the potential to be applied in a variety of fields and environments beyond simple vehicle operation, and changes in the technological and legal environment are enabling this transition (Akeb et al., 2018). The use of autonomous robots, such as delivery robots and patrol robots, is a concrete example of how autonomous technology can optimize city operations and respond to social and environmental challenges, demonstrating how technological advances can positively impact our daily lives and the future of cities. An example of autonomous delivery robots is shown in Figs. 1.9 and 1.10.

In conclusion, the proliferation of smartphones, the transition to an aging society, the development of SDVs, and the growing interest in sustainable development are the main factors driving the societal demand for autonomous robots. The development of autonomous mobile robots is gaining importance at the intersection of socio-cultural change and technological advancement, and is becoming an important vehicle for bringing positive change to our society. These changes show that the development of autonomous mobile robotics is more than just a technological achievement, but has the potential to contribute to solving social, economic, and environmental problems. Autonomous robots are expected to play an important role in moving toward a better future.

Fig. 1.9 Starship autonomous delivery robots (*Source* Starship technology)

Fig. 1.10 Autonomous outdoor security robot in operation in the Times Square subway (*Source* Jefferson Siegel for The New York Times)

1.3 Historical Context and Evolution of Autonomous Driving Technology

Autonomous driving has a unified definition in the automotive industry and can best be described as "the ability of a car to drive partially or completely on its own with limited or no human intervention" (George et al., 2021). It is also considered to be one of the most disruptive innovations likely to occur in the next few years, and is expected to have enormous societal impact in all sorts of areas (Rosenzweig & Bartl, 2015).

The history of autonomous driving technology began in the 1920s, with the introduction of the world's first radio-controlled car, the Linriccan Wonder, in 1926, marking the beginning of a journey of technological advancement. In the 1980s, the technology was further advanced with the introduction of the vision-guided Mercedes-Benz robotic van, and since then, the focus has been on vision-guided

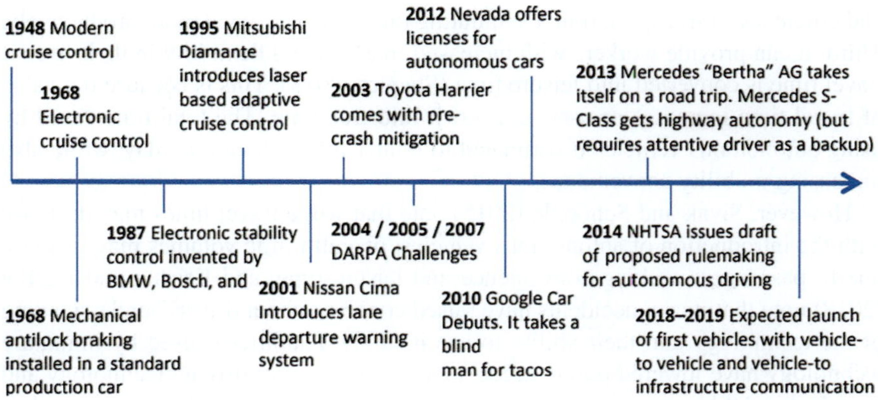

Fig. 1.11 Sixty-five years of automotive steady stages (*Source* Rosenzweig & Bartl, 2015)

systems using lidar, radar, GPS, and computer vision. These technologies have evolved into autonomous driving technologies in modern cars, such as adaptive cruise control, lane parking, and steer assist (Bimbraw, 2015), and newer models are adding new autonomous driving features such as information technology, communication, and data analysis and storage (Bimbraw, 2015). The evolution of automotive technology over time is shown in Fig. 1.11.

Since 2016, the Society of Automotive Engineers International (SAE International) classification of driving automation stages of autonomous driving has been used as a global standard. The stages are divided into six levels (0–5), with autonomous vehicles defined as self-driving starting at level 3. Currently, most commercial vehicles only include Level 1 to Level 2 autonomous driving capabilities that require constant driver attention and control due to the limitations and high cost of available sensors. The autonomous driving features in these vehicles typically consist of emergency braking, blind spot detection, and lane keeping (Van Brummelen et al., 2018).

The latest advances in autonomous driving technology include hardware innovations from automakers like Tesla and Hyundai, and software development from tech giants like Google, NVIDIA, and Intel. These companies are advancing the sensors, cameras, and data technology in autonomous vehicles, and developing advanced software that uses artificial intelligence and machine learning algorithms to enable vehicles to make their own decisions and navigate complex road conditions. These technological advances are not limited to autonomous vehicles, but also spill over into the field of robotics, with potential applications in delivery, cleaning, surveillance, and more.

Autonomous driving technology could have societal implications, and the potential impacts and benefits are somewhat predictable. First, it could increase the mobility of older adults, which could have a significant impact on the quality of life and productivity of older adults, who make up a significant portion of the population (Yurtsever et al., 2020). Second, autonomous systems can improve the safety and efficiency of transportation. Commuting time can be utilized more productively. Third, it can provide workers with increased freedom and flexibility in their lives if travel time is converted into leisure time (Shapiro, 2018). This is because the value of travel time increases as users convert it into free time. The total time saved by using autonomous vehicles is estimated to total 1 billion hours per day, while also increasing mobility convenience.

However, Sivak and Schoettle (2015) note that while travel times may decrease with the introduction of autonomous vehicles, overall traffic volumes may increase due to passengers seeking convenience and leisure time, and Krisher and Durbin (2018) note that recent accidents have raised concerns about drivers' understanding of the technology and their ability to use it safely. Fatalities caused by immature technology have lowered public acceptance of autonomous driving technology and raised ethical dilemmas about autonomous driving (Yurtsever et al., 2020). It will be necessary to develop the technology to resolve these situations and identify ways to resolve the ethical dilemmas in parallel with technological advancements.

Even if the transition to mobility with autonomous technology takes longer than expected, it will contribute to improving the sustainability of local urban transportation systems, and the spatial context in which autonomous technology development should be planned to prepare for the diffusion of autonomous driving will need to be considered as it scales from single municipalities to entire metropolitan areas (Yurtsever et al., 2020).

1.4 Expected Effects and Role of Autonomous Robots in the Future Society

Autonomous robots have recently gained popularity in modern society, and the global market for these robots is expected to grow significantly over the next two decades. Initially, robots were mostly industrial robots, but they are expanding into various fields such as medical robots (Zhang et al., 2021), logistics robots (karabegović et al., 2015), etc. Autonomous robots are being used to improve everyday activities and services, including healthcare (Takahashi et al., 2010; Calderon et al., 2015), logistics (Volos et al., 2013), cleaning (Chen et al., 2023), security checks (Wang et al., 2020), and more (Zghair & Al-Araji 2021).

You can see that these autonomous robots are being used to handle indoor tasks. However, in recent years, autonomous robots have expanded their spatial scope to include not only indoors but also outdoors, and autonomous robots have emerged in various environments. As laws and perceptions of pedestrian environments change, autonomous robots are being commercialized to drive on sidewalks (Akeb et al., 2018). Autonomous robots operating outdoors are increasingly being used as delivery robots to provide food delivery services (Gehrke et al., 2023), patrol robots for protection and patrolling (Šelek et al., 2023), and quarantine robots to prevent the

Table 1.5 Trends in pedestrian walkway autonomous robotics utilization and technology

Dust bot (Dustbot)	Starship technology (Starship Technologies)

Responsible for trash collection and cleaning sidewalks, utilizing laser and ultrasound scanners	Food delivery and last mile parcel delivery services, Level 4 autonomous driving
Ford&Agility Robotics (Ford)	ROBOTIS (ROBOTIS)

With self-driving vehicles and 'Digit' robots, the last mile parcel delivery service, walking around on two legs, equipped with lidar and stereo cameras	Provide various delivery services with indoor and outdoor self-driving robots, Currently operating in partnership with hotels and various organizations

spread of viruses (Mehta et al., 2023), and trends in the use and technology of various sidewalk autonomous robot services are summarized in Table 1.5.

Along with these technological advances, the introduction and proliferation of autonomous robots mentioned above has the potential to reshape cities and redefine the way we live in the future. The expansion of autonomous robots has also seen a marked increase in the private sector, and is driving innovation in many aspects of urban life. This means that technological advancements are going beyond mere convenience and are dramatically changing our daily lives.

References

Akeb, H., Moncef, B., & Durand, B. (2018). Building a collaborative solution in dense urban city settings to enhance parcel delivery: An effective crowd model in Paris. *Transportation Research Part E: Logistics and Transportation Review, 119*, 223–233.

Alatise, M. B., & Hancke, G. P. (2020). A review on challenges of autonomous mobile robot and sensor fusion methods. *IEEE Access, 8*, 39830–39846.

Aloisio et al. (2012). Next-generation image-guided citrus fruit picker. In: *International Conference on Technologies for Practical Robot Applications (TePRA)* (pp. 37–41).

Bertozzi, M., Broggi, A., & Fascioli, A. (2000). Vision-based intelligent vehicles: State of the art and perspectives. *Robotics and Autonomous Systems, 32*(1), 1–16.

Bimbraw, K. (2015, July). Autonomous cars: Past, present and future a review of the developments in the last century, the present scenario and the expected future of autonomous vehicle technology. In *2015 12th International Conference on Informatics in Control, Automation and Robotics (ICINCO)* (Vol. 1, pp. 191–198). IEEE.

Calderon, C. A. A., Mohan, E. R., & Ng, B. S. (2015). Development of a hospital mobile platform for logistics tasks. *Digital Communications and Networks, 1*(2), 102–111.

Cebollada, S., Payá, L., Flores, M., Peidró, A., & Reinoso, O. (2021). A state-of-the-art review on mobile robotics tasks using artificial intelligence and visual data. *Expert Systems with Applications, 167*, 114195.

Chen, J., Zhu, Y., Li, Z., Chen, X., Chen, X., Huang, S., … Zhuo, Y. (2024). Global impact of population aging on vision loss prevalence: A population-based study. *Global Transitions, 6*, 28–36.

Chen, L. B., Huang, X. R., Chen, W. H., Pai, W. Y., Huang, G. Z., & Wang, W. C. (2023). Design and implementation of an artificial intelligence of things-based autonomous mobile robot system for cleaning garbage. *IEEE Sensors Journal.*

Deb, S., Rahman, M. M., Strawderman, L. J., & Garrison, T. M. (2018). Pedestrians' receptivity toward fully automated vehicles: Research review and roadmap for future research. *IEEE Transactions on Human-Machine Systems, 48*(3), 279–290.

Gebbers, R., & Adamchuk, V. I. (2010). Precision agriculture and food security. *Science, 327*(5967), 828–831.

Gehrke, S. R., Phair, C. D., Russo, B. J., & Smaglik, E. J. (2023). Observed sidewalk autonomous delivery robot interactions with pedestrians and bicyclists. *Transportation Research Interdisciplinary Perspectives, 18*, 100789.

George, D. et al. (2021). A path of structural transformation for the automotive and insurance industries toward autonomous vehicles. In: *Autonomous vehicles.* Elsevier

Huang, B., Huan, Y., Xu, L. D., Zheng, L., & Zou, Z. (2019). Automated trading systems statistical and machine learning methods and hardware implementation: A survey. *Enterprise Information Systems, 13*(1), 132–144.

Karabegović, I., Karabegović, E., Mahmić, M., Husak, E. J. A. I. P. E. (2015). The application of service robots for logistics in manufacturing processes. In: *Advances in Production Engineering & Management, 10*(4).

Khan, M. A. (2021). Intelligent environment enabling autonomous driving. *IEEE Access, 9*, 32997–33017.

Krisher, T., & Durbin, D.-A. (2018). *Tesla update halts automatic steering if driver inattentive.* PHYSORG.https://phys.org/news/2016-09-tesla-halts-automatic-driver-inattentive.html

Mehta, I., Hsueh, H. Y., Taghipour, S., Li, W., & Saeedi, S. (2023). UV disinfection robots: A review. *Robotics and Autonomous Systems, 161*, 104332.

Mirnig, N., Perterer, N., Stollnberger, G., & Tscheligi, M. (2017). Three strategies for autonomous car-to-pedestrian communication: A survival guide. In: *Proceedings of Companion 2017 ACM/IEEE International Conference on Human-Robot Interaction* (pp. 209–210).

Park, J. G. (2019). *From perceptron to CNN, a history of deep learning.* The Korea Institute of Technology Press. https://times.kaist.ac.kr/news/articleView.html?idxno=4675

Pettigrew, S., Talati, Z., & Norman, R. (2018). The health benefits of autonomous vehicles: Public awareness and receptivity in Australia. *Australian and New Zealand Journal of Public Health, 42*(5), 480–483.

Rahmadian, R., & Widyartono, M. (2020, October). Autonomous robotic in agriculture: a review. In *2020 Third International Conference on Vocational Education and Electrical Engineering (ICVEE)* (pp. 1–6). IEEE.

Rosenzweig, J., & Bartl, M. (2015). A review and analysis of literature on autonomous driving. *E-Journal Making-of Innovation*, 1–57.

Ross, P. E. (2014). Robot, you can drive my car. *IEEE Spectrum, 51*(6), 60–90.

Rouff, C. (2012). M. Hinchey (Ed.), *Experience from the DARPA urban challenge*. Springer.

Šelek, A., Seder, M., & Petrović, I. (2023). Smooth autonomous patrolling for a differential-drive mobile robot in dynamic environments. *Sensors, 23*(17), 7421.

Shapiro, A. (2018). Between autonomy and control: Strategies of arbitrage in the "on-demand" economy. *New Media & Society, 20*(8), 2954–2971.

Shishehgar, M., Kerr, D., & Blake, J. (2018). A systematic review of research into how robotic technology can help older people. *Smart Health, 7*, 1–18.

Sivak, M., & Schoettle, B. (2015). *Road safety with self-driving vehicles: General limitations and road sharing with conventional vehicles*. University of Michigan.

Takahashi, M., Suzuki, T., Shitamoto, H., Moriguchi, T., & Yoshida, K. (2010). Developing a mobile robot for transport applications in the hospital domain. *Robotics and Autonomous Systems, 58*(7), 889–899.

Transportation, U. S. D. O. (2016). *Federal automated vehicles policy*.

Vaigandla, K. K., Bolla, S., & Karne, R. (2021). A survey on future generation wireless communications-6G: requirements, technologies, challenges and applications. *International Journal, 10*(5).

Van Brummelen, J., O'Brien, M., Gruyer, D., & Najjaran, H. (2018). Autonomous vehicle perception: The technology of today and tomorrow. *Transportation Research Part c: Emerging Technologies, 89*, 384–406.

Wang, C., Tok, Y., Poolat, R., Chattopadhyay, S., Mohan, R. E. (2020). How to Secure Autonomous Mobile Robots? An Approach with Fuzzing, Detection and Mitigation. *Journal of Systems Architecture, 112*. 101838. https://doi.org/10.1016/j.sysarc.2020.101838.

Yurtsever, E., Lambert, J., Carballo, A., & Takeda, K. (2020). A survey of autonomous driving: Common practices and emerging technologies. *IEEE Access, 8*, 58443–58469.

Zghair, N. A. K. & Al-Araji, A. (2021). A one decade survey of autonomous mobile robot systems. *International Journal of Electrical and Computer Engineering, 11*. 4891–4906. https://doi.org/10.11591/ijece.v11i6.pp4891-4906.

Zhang, C., & Lu, Y. (2021). Study on artificial intelligence: The state of the art and future prospects. *Journal of Industrial Information Integration, 23*, 100224.

Zhang, S., Klein, D. A., Bauckhage, C., & Cremers, A. B. (2016). Fast moving pedestrian detection based on motion segmentation and new motion features. *Multimedia Tools and Applications, 75*(11), 6263–6282.

Zhang, W., Li, H., Cui, L., Li, H., Zhang, X., Fang, S., & Zhang, Q. (2021). Research progress and development trend of surgical robot and surgical instrument arm. *The International Journal of Medical Robotics and Computer Assisted Surgery, 17*(5), e2309.

Chapter 2
Advances in Autonomous Driving Robots: Past, Present and Future

The field of autonomous driving has rapidly evolved, driven by technological breakthroughs and an increasing demand for safer, more efficient urban mobility solutions. The transition from early theoretical concepts to real-world applications has been marked by significant advancements in artificial intelligence, sensor technology, and system integration. This chapter provides an in-depth exploration of the evolution of autonomous robots, examining their historical development, current applications, and future potential.

Autonomous driving technology is more than just a technical achievement—it is a response to pressing societal challenges. Urban congestion, environmental pollution, and the growing need for accessible transportation have all accelerated the development of innovative mobility solutions. Autonomous vehicles are increasingly seen as a crucial tool in addressing these challenges, offering the promise of reducing accidents caused by human error, optimizing transportation systems, and contributing to a more sustainable future. However, these technological advancements also introduce complex challenges, including ethical dilemmas, regulatory hurdles, and the impact on traditional labor markets. This chapter aims to present a balanced analysis of both the opportunities and obstacles associated with autonomous mobility, providing insights into its transformative potential and the challenges that must be addressed for widespread adoption.

Section 2.1 traces the evolution of autonomous driving from the 1980s to the present, highlighting key experiments and technological milestones that have laid the foundation for today's sophisticated autonomous systems. This section reviews landmark projects and innovations that have shaped the current landscape of autonomous vehicle technology.

Following this, Sect. 2.2 shifts focus to the present, exploring the numerous real-world applications and recent advancements in autonomous systems. It examines how autonomous vehicles are currently being deployed across various settings—from bustling urban centers to specialized industrial environments—demonstrating their substantial impact on modern mobility and public services.

M. Choi and S. Lee, *Intelligent Moving Cities: Technological Leap and Social Integration of Autonomous Mobile Robots*,
https://doi.org/10.1007/978-981-96-8864-7_2

Finally, Sect. 2.3 looks toward the future of autonomous driving, discussing emerging technologies, potential new applications, and the societal and ethical challenges that must be carefully navigated to fully unlock the potential of this transformative technology.

Overall, this chapter provides a comprehensive perspective on the evolution of autonomous driving by analyzing its technical, practical, and ethical dimensions. By tracing its development from historical roots to current innovations and future prospects, we gain a clearer understanding of the profound impact that autonomous technologies can have on urban mobility, industrial efficiency, and public services.

Additionally, this chapter underscores the crucial need for collaboration among governments, technology companies, and various stakeholders to ensure that autonomous driving advances in a manner that benefits society, enhances public safety, and adheres to ethical standards.

As autonomous driving technology continues to mature, the establishment of robust infrastructure, regulatory frameworks, and public acceptance will be key to its successful integration. The journey of autonomous vehicles—from conceptual research to real-world deployment—is not just about technological innovation; it is about ensuring that these systems seamlessly integrate into the fabric of our cities and communities. This chapter lays the groundwork for understanding how autonomous driving has evolved, its current impact, and how it may shape the future by fostering a more connected, efficient, and equitable society.

2.1 The Evolution of Autonomous Driving: From Past to Present

Autonomous driving technology has undergone significant changes over the past few decades, evolving from theoretical experiments to advanced practical applications. The journey of autonomous driving robots is a testament to the evolution of artificial intelligence, sensor technology, and human ambition to create more efficient and sustainable urban mobility systems. In this section, we take an in-depth look at how autonomous driving has evolved over time, looking back at key milestones and technological advances that paved the way for today's sophisticated systems.

2.1.1 The Birth of Autonomous Driving: Early Developments (1980s–1990s)

The origins of autonomous driving can be traced back to early research initiatives in the 1980s and 1990s. During this time, much of the research was exploratory, focused on building basic machine perception and developing navigation systems that could mimic human driving abilities. One of the earliest examples is the Navlab project developed at Carnegie Mellon University, which involved a series of autonomous vehicles with rudimentary sensors and onboard computing, laying the groundwork for understanding how vehicles interpret and navigate their surroundings (Jochem, 1995).

Another important development of this period was the introduction of the Prometheus project in Europe, led by Daimler Benz. Prometheus, which stands for "Program for European Transport with Maximum Efficiency and Unprecedented Safety," represented an ambitious leap forward in the exploration of automated vehicle technology. With a strong focus on safety, the project was one of the first to experiment with onboard cameras and computer vision systems for road lane detection and obstacle recognition. Advances made in the 1980s and 1990s focused primarily on the development of sensor-based perception systems, which were an important step towards truly autonomous mobility.

2.1.2 The Rise of AI and Sensor Fusion (2000s–2010s)

The 2000s marked a turning point in the evolution of autonomous driving: rapid advances in artificial intelligence and machine learning, along with improvements in sensor technology, enabled many of the challenges faced by early systems to be overcome. The Defense Advanced Research Projects Agency (DARPA) played an important role in advancing autonomous driving technology by hosting the first Grand Challenge in 2004. Although no team completed the first course, the lessons learned from the competition and the enthusiasm for AI-powered vehicles sparked renewed interest in innovation (Fig. 2.1).

Fig. 2.1 The grand challenge (*Source*: https://www.darpa.mil/about/innovation-timeline)

The DARPA Urban Challenge in 2007 took the competition to the next level, shifting the testing environment to an urban setting where vehicles had to navigate complex traffic scenarios, including intersections, roundabouts, and real-world obstacles. The competition introduced the need for sensor fusion, which combines data from multiple sensors such as lidar, radar, and cameras to give vehicles a more comprehensive understanding of their surroundings (Fig. 2.2). Advances in AI algorithms have also enabled vehicles to make real-time decisions, improving their ability to adapt to unpredictable environments.

During this decade, companies like Google (now Waymo) entered the autonomous vehicles market and pushed the limits of autonomous vehicles. In 2010, Google's autonomous driving car project demonstrated a new level of reliability for autonomous systems, proving that cars could navigate city traffic with great precision. These advances were largely due to improvements in machine learning technology, particularly neural networks, which enabled vehicles to recognize and classify objects with greater accuracy. Lidar technology has become an integral part of autonomous driving, providing HD 3D maps that are essential for safe and efficient travel (Waymo) (Fig. 2.3).

Fig. 2.2 DARPA urban challenge (*Source* https://www.darpa.mil/about/innovation-timeline)

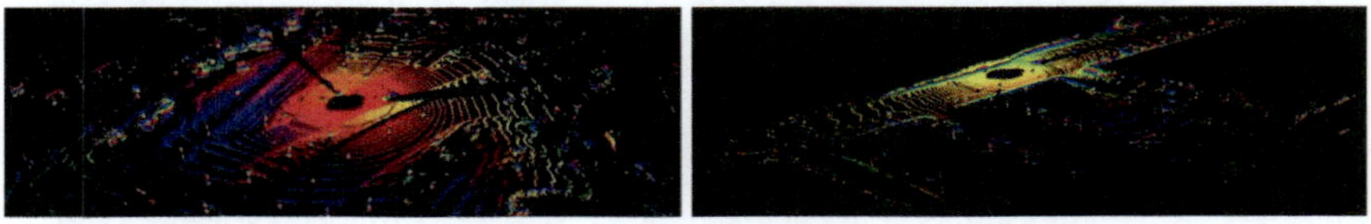

Fig. 2.3 Autonomous technology: 3D point clouds captured by LiDAR (*Source* Liu et al., 2023)

2.1.3 Current Technology Landscape: Advanced Models and Key Milestones (2010s–2020s)

In the 2010s and beyond, autonomous driving transitioned from experimental projects to commercial deployment. The focus was on improving the technology to achieve higher levels of autonomy, from Level 1 (driver assistance) to Level 5 (full automation). This period saw significant innovations, including advances in deep learning, reinforcement learning, and Vehicle-to-Vehicle (V2X) communication technologies.

Companies like Tesla, Waymo, and Baido have developed sophisticated autonomous driving platforms that harness the power of AI for decision-making and route planning. Tesla's Autopilot and fully autonomous driving capabilities are among the most controversial and ambitious efforts to bring autonomy to everyday vehicles. Tesla's approach, which relies on vision-based navigation rather than lidar, contrasts with Waymo's, which uses a combination of lidar, radar, and cameras to achieve a comprehensive perception of the environment.

The role of HD (high-definition) maps has also become key to achieving advanced levels of autonomous driving. Companies like Nvidia are also entering this space, providing specialized hardware such as custom GPUs for real-time data processing that is essential for the split-second decision-making required in urban environments. These hardware advances are critical because they allow autonomous vehicles to quickly process vast amounts of data and make more informed navigation decisions (Jebamikyous & Kashef, 2022).

An important milestone in this period is the commercialization of autonomous driving services. In 2020, Waymo launched Waymo One, the world's first fully autonomous ride-hailing service available to the public in parts of Phoenix, Arizona, a significant achievement that demonstrated that autonomous driving vehicles can operate safely in real-world environments without human intervention. In addition, autonomous driving delivery robots from companies like Nuro and Starship Technologies are becoming commonplace in some urban areas, reflecting the expansion of autonomous driving beyond personal vehicles into a variety of mobility solutions.

The current landscape of autonomous driving is characterized by collaboration between automakers, technology companies, and governments, and the introduction of 5G networks is playing a key role in enabling real-time communication between vehicles and infrastructure to support advanced features such as swarming and cooperative navigation. The field is currently on the verge of widespread deployment, with numerous pilot projects underway in major cities around the world.

From the 1980s to the present, the evolution of autonomous driving has been driven by a combination of technological innovation, cross-disciplinary collaboration, and societal demand for safer and more efficient transportation systems. From early experimental models with basic sensor capabilities to today's sophisticated AI-powered autonomous systems, this journey reflects both challenges and achievements to realize the vision of autonomous mobility. The next phase will build on this foundation and push the boundaries of what autonomous robots can do in urban environments.

2.2 Current Applications and Achievements of Autonomous Driving Technology

As autonomous driving technology has moved from the lab to the real world, it has a variety of applications that have reshaped the way we perceive mobility in cities. This section explores current real-world applications and achievements that have demonstrated the feasibility and benefits of autonomous systems. A more detailed discussion of current examples will be covered in Chap. 6.

2.2.1 Autonomous Driving in Urban Environments

One of the most notable applications of autonomous technology is in urban mobility, where autonomous vehicles are being used to enhance public and private transportation options. For example, autonomous shuttles are currently being deployed in several cities to provide last-mile connections. These shuttles are designed to operate in heavily pedestrianized areas and provide safe and reliable transportation to bridge the gap between major transit hubs and local destinations.

Examples of autonomous shuttles include those from companies like Navya and EasyMile, which are operating in pilot programs in several cities in Europe and North America (Fig. 2.4). These shuttles aim to improve accessibility while reducing traffic congestion and emissions. Utilizing a combination of lidar, GPS, and cameras, these vehicles can safely navigate complex roads, avoid obstacles, and ensure a smooth ride for passengers.

In addition to shuttles, autonomous delivery robots are also finding their place in urban environments. Companies like Starship Technologies and Nuro have developed small autonomous robots designed for short-distance deliveries (Fig. 2.5). These robots are particularly effective at delivering groceries, meals, and other small packages, reducing the need for traditional delivery vehicles, and contributing to lower

Fig. 2.4 Autonomous shuttles (*Source* PPPR, 2020)

Fig. 2.5 Nuro's driverless delivery service (*Source* Das, 2021)

emissions. The COVID-19 pandemic has accelerated the adoption of these robots, highlighting the need for contactless delivery solutions.

2.2.2 Industrial and Specialty Applications

Beyond urban mobility, autonomous technology is also making significant progress in industrial and specialty environments. Autonomous vehicles are being deployed in airports, warehouses, factories, and other controlled environments where they can perform repetitive tasks with high precision and efficiency. In these environments, autonomous robots are used for tasks such as cargo handling, inventory management, and transporting materials between facilities.

At airports, for example, autonomous tugs and tractors are used to move baggage and cargo between terminals, increasing operational efficiency and reducing labor costs. In warehouses, companies like Amazon and Alibaba are employing autonomous robots to streamline inventory management, allowing them to pick, pack, and ship products with minimal human intervention (Fig. 2.6). These specialized applications highlight the versatility of autonomous driving technology and its potential to revolutionize industries by increasing efficiency and reducing human error.

Fig. 2.6 Alibaba's autonomous robots in Warehouse (*Source* Prawira, 2023)

2.2.3 Autonomous Driving in Public Services

Another important application of autonomous technology is in public services. Autonomous street sweeping vehicles and security patrol robots are being used in cities to maintain cleanliness and safety. These robots are designed to autonomously navigate city streets and perform a given task while avoiding pedestrians and other obstacles. For example, Singapore is testing autonomous street sweeping robots to clean public roads, providing a cost-effective solution to urban sanitation issues.

Security robots developed by Knightscope are also being deployed in public spaces such as shopping malls, parking lots, and business districts. These robots are equipped with cameras, sensors, and communication systems to detect suspicious activity and report it to authorities in real time. By providing continuous security, these robots enhance public safety while freeing up human resources to focus on more complex tasks.

2.3 Future Prospects and Developments in Autonomous Driving Technology

As autonomous driving technology continues to evolve, the future holds tremendous potential for new innovations and expanded applications. This section explores the future outlook for autonomous driving, focusing on emerging technologies, potential use cases, and challenges that need to be addressed to realize the full potential of autonomous systems.

2.3.1 Technological Innovations: AI, Edge Computing, and Beyond

The next phase of autonomous driving technology will be driven by advances in artificial intelligence, edge computing, and communication technologies. AI will continue to play a pivotal role in enhancing the capabilities of autonomous vehicles, with a particular focus on improving real-time decision-making and perception accuracy. The development of more sophisticated deep learning models and reinforcement learning algorithms will enable autonomous systems to better understand complex urban environments, predict pedestrian and vehicle behavior, and make informed decisions.

Edge computing is also expected to revolutionize autonomous driving by bringing data processing closer to the source. Unlike cloud computing, which sends data to a centralized server for processing, edge computing can process data locally within the vehicle, reducing latency for faster response times, which is critical to the safety and efficiency of autonomous vehicles. The integration of edge computing and AI can lead to more robust and reliable autonomous driving systems, capable of handling a wider variety of scenarios in real time (Fig. 2.7) (Gupta et al., 2023).

Advances in sensor technology will also continue to improve the perception capabilities of autonomous vehicles. High-resolution cameras, next-generation LIDAR, and radar systems will provide a more detailed and accurate representation of the environment, enabling better obstacle detection and path planning. The combination of these sensors with AI and edge computing will result in more resilient and adaptive autonomous systems.

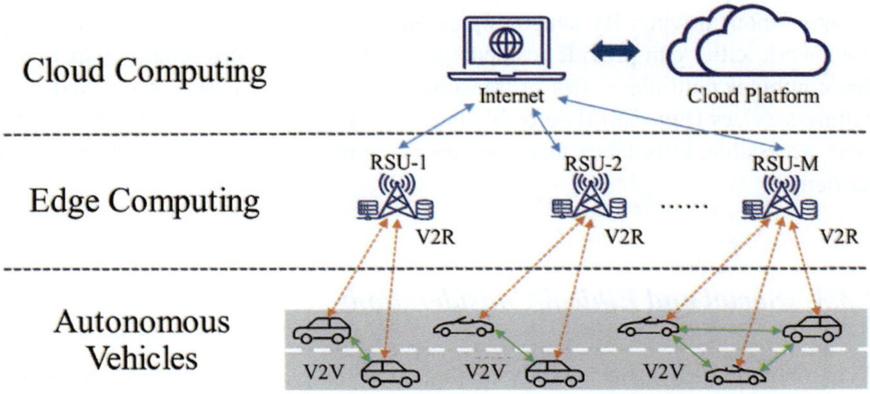

Fig. 2.7 Edge computing-based connected autonomous Vehicles (*Source* Gupta et al., 2023)

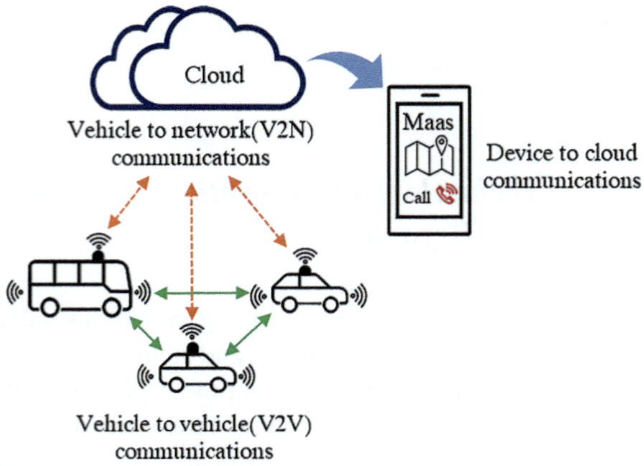

Fig. 2.8 Architecture for connected autonomous vehicles in MaaS (*Source* Lee & Madnick, 2021)

2.3.2 Application Expansion and City Integration

The future of autonomous driving is not limited to personal vehicles but is expanding into a variety of urban mobility solutions. One of the most promising areas is the integration of autonomous vehicles into public transportation networks. Autonomous buses and shuttles can provide efficient, on-demand transportation services, reducing the need for private vehicle ownership and alleviating traffic congestion in urban centers (Iclodean et al., 2020). This can contribute to a sustainable urban environment by reducing emissions and increasing the utilization of public space.

Autonomous technology also has the potential to revolutionize logistics and delivery services. Autonomous trucks can be utilized for long-distance cargo transportation, reducing reliance on human drivers and addressing the driver shortage in the logistics industry (Hasiri & Kermanshah, 2024). In urban areas, autonomous delivery robots and drones can be used for last-mile delivery to deliver goods quickly and efficiently with minimal environmental impact (Brunner et al., 2019).

The concept of MaaS (mobility-as-a-service) will also benefit from advances in autonomous driving. By integrating autonomous vehicles into a shared mobility framework, cities can provide residents with a seamless transportation experience that combines multiple modes of transportation, including buses, trains, and ride-hailing services (Fig. 2.8) (Lee & Madnick, 2021). This will make urban mobility more accessible, affordable, and convenient, improving the quality of life for city residents.

2.3.3 Social and Ethical Considerations

As autonomous driving technology continues to evolve, it is essential to consider the social and ethical implications of its adoption. One of the main concerns is safety. autonomous driving vehicles have the potential to reduce accidents caused by human error, but we need to ensure that these systems are reliable and can handle unexpected situations. Rigorous testing, validation, and certification procedures are needed to build public trust and ensure the safe operation of autonomous vehicles (Rhim et al., 2021).

Another important consideration is the impact on employment. Widespread adoption of autonomous driving technology could result in the loss of driver jobs in industries such as transportation and logistics. Policymakers will need to address these issues by developing strategies to retrain workers and create new jobs in emerging sectors related to autonomous driving technology.

Ethical considerations also play an important role in the development of autonomous systems. Autonomous vehicles must be programmed to make ethical decisions in situations where human lives are at stake. The challenge is to develop algorithms that can balance safety, legality, and moral considerations in complex scenarios. The development of ethical frameworks and guidelines is necessary to ensure that autonomous systems are in line with societal values and expectations (Karmouskos, 2018).

2.3.4 Future Challenges and Opportunities

Despite the significant progress made in autonomous driving technology, several challenges need to be addressed to realize its full potential. One of the key challenges is regulation. Governments around the world are grappling with how to regulate the deployment of autonomous vehicles while balancing the need for innovation with public safety. Developing a consistent regulatory framework that can accommodate the rapid pace of technological change is critical to the widespread adoption of autonomous driving.

Another challenge is infrastructure. For autonomous vehicles to operate effectively, they need a robust and well-maintained infrastructure. This includes high-quality road surfaces, reliable communication networks, and smart traffic management systems. Cities will need to invest in upgrading their infrastructure to support the deployment of autonomous vehicles and ensure they operate safely and efficiently (Fig. 2.9).

Meanwhile, the future offers numerous opportunities for autonomous driving technology. The continued development of smart cities with integrated sensors and communication systems will create an environment conducive to the deployment of autonomous vehicles, which will be able to utilize data from connected infrastructure to drive more safely and efficiently, providing a better experience for users.

The evolution of autonomous driving technology is ongoing, and the future is bright. By addressing the challenges and capitalizing on the opportunities, autonomous vehicles have the potential to revolutionize urban mobility, making cities more livable, sustainable, and connected. The future of autonomous driving lies in the seamless integration of technology, infrastructure, and societal needs to usher in a new era of mobility.

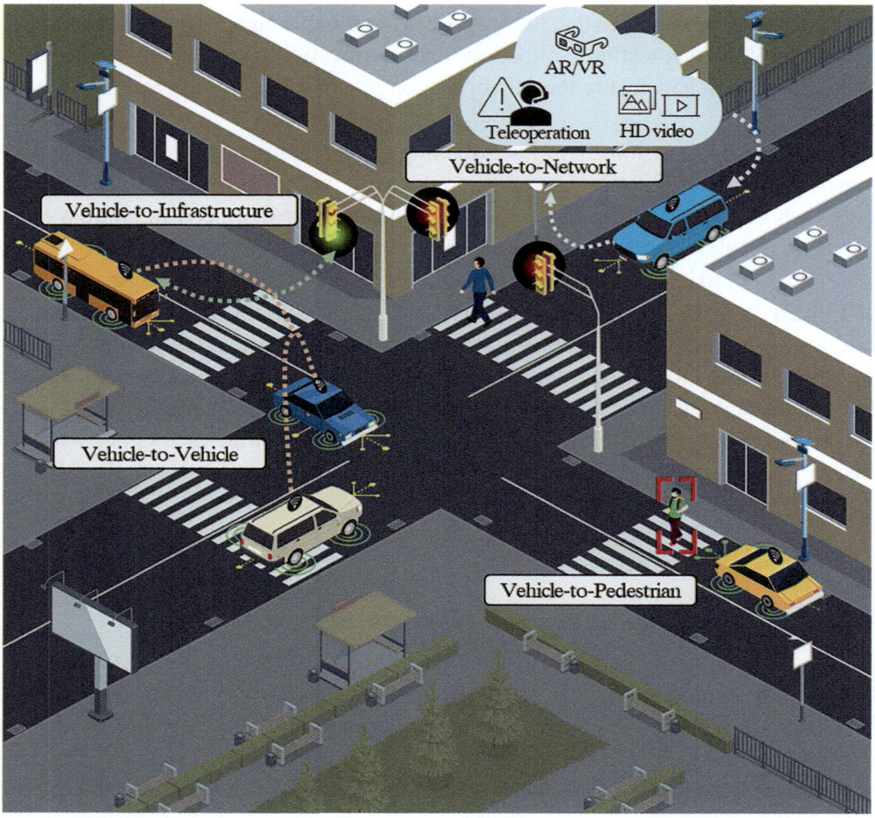

Fig. 2.9 Smart cities communications architecture with autonomous vehicles (*Source* Talib et al., 2018)

References

Brunner, G., Szebedy, B., Tanner, S., & Wattenhofer, R. (2019, June). The urban last mile problem: Autonomous drone delivery to your balcony. In *2019 International Conference on Unmanned Aircraft Systems (ICUAS)* (pp. 1005–1012). IEEE.

Das, S., Wei, Z., & Ravuri, V. (2021). *Safety and operations of automated delivery vehicles: A scoping review.*

Gupta, S., Maple, C., & Passerone, R. (2023). *An investigation of cyber-attacks and security mechanisms for connected and autonomous vehicles.* IEEE Access.

Hasiri, A., & Kermanshah, A. (2024). Exploring the role of autonomous trucks in addressing challenges within the trucking industry: A comprehensive review. *Systems, 12*(9), 320.

https://media.mercedes-benz.com/article/b7bf109a-260a-4031-91b4-9d32830dfd49

https://waymo.com/waymo-driver/

https://www.darpa.mil/about-us/timeline/darpa-urban-challenge

Iclodean, C., Cordos, N., & Varga, B. O. (2020). Autonomous shuttle bus for public transportation: A review. *Energies, 13*(11), 2917.

Jebamikyous, H. H., & Kashef, R. (2022). Autonomous vehicles perception (AVP) using deep learning: Modeling, assessment, and challenges. *IEEE Access, 10*, 10523–10535.

Jochem, T., Pomerleau, D., Kumar, B., & Armstrong, J. (1995, September). PANS: A portable navigation platform. In *Proceedings of the Intelligent Vehicles' 95. Symposium* (pp. 107–112). IEEE.

Karnouskos, S. (2018). Self-driving car acceptance and the role of ethics. *IEEE Transactions on Engineering Management, 67*(2), 252–265.

Lee, C. W., & Madnick, S. (2021). Cybersafety approach to cybersecurity analysis and mitigation for mobility-as-a-service and internet of vehicles. *Electronics, 10*(10), 1220.

Liu, H., Wu, C., & Wang, H. (2023). Real time object detection using LiDAR and camera fusion for autonomous driving. *Scientific Reports, 13*(1), 8056.

PPPR, N. L. (2020). *Campus automated shuttle service deployment initiative.*

Prawira, I. F. A., Habbe, A. H., Muda, I., Hasibuan, R. M., & Umbrajkaar, A. (2023, April). Robot as staff: Robot for Alibaba e-commerce warehouse pro-cess. In *2023 International Conference on Inventive Computation Technologies* (ICICT) (pp. 1619–1623). IEEE.

Rhim, J., Lee, J. H., Chen, M., & Lim, A. (2021). A deeper look at autonomous vehicle ethics: An integrative ethical decision-making framework to explain moral pluralism. *Frontiers in Robotics and AI, 8*, 632394.

Talib, M. S., Hassan, A., Hussin, B., Mohammed, A. A. J., Hassan, A. A., & Mutlag, A. A. (2018). Vehicular ad hoc network for intelligent transport system: A review. *International Journal of Engineering and Technology, 7*.

Chapter 3
Technical Considerations for Autonomous Navigation and Route Navigation

As autonomous mobility continues to evolve, the ability of robots to navigate complex urban environments has become a crucial aspect of modern city planning. This chapter explores the technical considerations of autonomous driving and route optimization in pedestrian-dense urban areas, focusing on the intersection of technology, efficiency, and real-time decision-making. Cities present unique challenges for autonomous mobile robots, including unpredictable human behavior, dynamically changing obstacles, and the need for efficient energy consumption.

Navigating such environments is inherently complex, requiring both pre-planned strategies and real-time adaptability. Unlike static navigation systems, autonomous mobile robots must continuously interpret sensor data, make rapid decisions, and adjust their routes dynamically. This chapter delves into these challenges, starting with fundamental path-finding algorithms and progressing to more advanced adaptive navigation techniques.

A key focus is on understanding and adapting navigation algorithms such as Dijkstra, A*, and D* for urban scenarios. Additionally, this chapter examines the impact of robot-pedestrian interactions on navigation strategies and how to optimize these interactions for both efficiency and safety. Energy efficiency is another critical factor, as battery capacity limits the operational time of autonomous mobile robots. Therefore, balancing energy-efficient path planning with real-time responsiveness is a major consideration.

This chapter is divided into three main sections.

Section 3.1 introduces the fundamental principles of navigation systems, covering basic path-finding algorithms and the concepts of global and local path planning.

Section 3.2 explores advanced navigation strategies, including modifications to existing algorithms to better handle pedestrian environments. This section also discusses solutions to human–robot interaction challenges.

Finally, Sect. 3.3 highlights the importance of real-time route planning and energy efficiency, introducing methods such as reinforcement learning to enhance operational sustainability.

M. Choi and S. Lee, *Intelligent Moving Cities: Technological Leap and Social Integration of Autonomous Mobile Robots*,
https://doi.org/10.1007/978-981-96-8864-7_3

By systematically addressing these aspects, this chapter provides a comprehensive overview of the technical challenges and solutions involved in autonomous navigation in urban environments. It bridges the gap between theoretical algorithm development and real-world implementation, offering insights into how autonomous systems can be seamlessly integrated into pedestrian-centric cities to enhance urban mobility.

3.1 The Basics of How to Build a Navigation System

The navigation system is the core of all autonomous mobile robot platforms, providing the essential functions for moving in complex environments. For autonomous mobile robots operating in pedestrian-oriented cities, efficient navigation is of paramount importance as they must coexist with various urban elements such as infrastructure, dynamic obstacles, and pedestrians. This section aims to provide a comprehensive introduction to the basic principles of navigation systems, providing a basic understanding of the different algorithms that are important for static and dynamic path finding in a variety of scenarios, ranging from general navigation systems to sophisticated autonomous mobile robotics.

3.1.1 Overview of Navigation and Route Finding Algorithms

The basis of all navigation systems starts with the route-finding algorithm, which determines the optimal route from the starting point to the destination. Over the years, various algorithms have been developed and applied to solve the requirements of autonomous navigation for vehicles, drones, and autonomous mobile robots. Among the most widely used algorithms are **Dijkstra's algorithm** (Fig. 3.1), **A*, and D* (dynamic A*)** (Figs. 3.2 and 3.3). These algorithms have played an important role in the development of basic navigation and advanced autonomous driving systems.

1. **Dijkstra's algorithm** is a graph search algorithm that finds the shortest path between nodes in a graph, making it ideal for scenarios where all edge costs are non-negative. While this algorithm guarantees to find the shortest path, it can be computationally expensive, especially in large environments. It provides a good foundation for understanding shortest path algorithms in navigation systems.
2. **A***, an improvement on Dijkstra, uses a heuristic to prioritize certain paths and focus on the most promising paths first, greatly reducing computation time. A* is very effective in a static environment where obstacles do not change, making it suitable for a wide range of robotic applications, including ground and aerial vehicles. It balances optimality and computational efficiency by using a cost function that estimates the distance to the target.

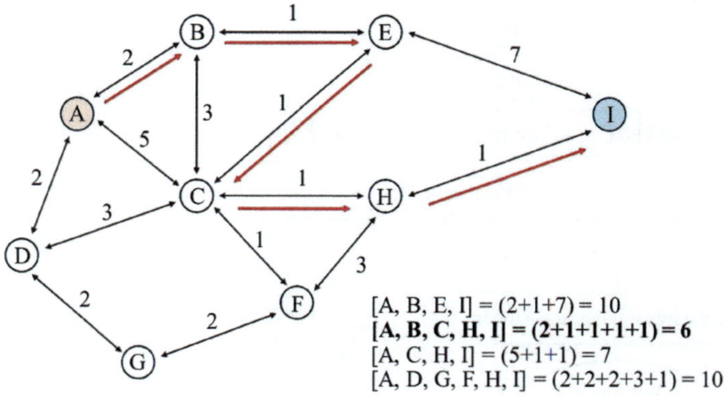

Fig. 3.1 Illustration of Dijkstra's algorithm (*Source* Rehman et al., 2019)

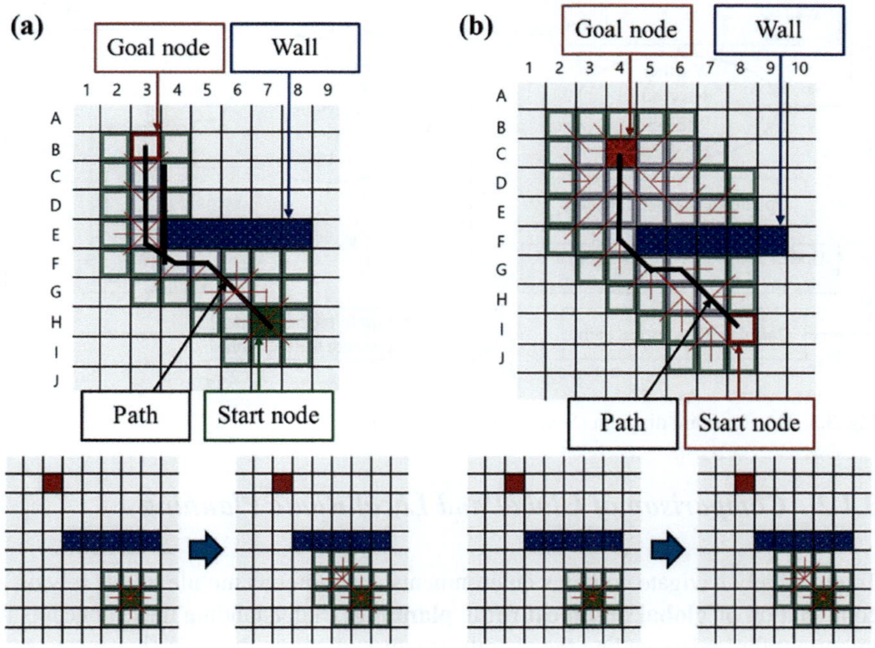

Fig. 3.2 Illustration of algorithm: **a** A* algorithm, **b** D* algorithm (*Source* Kim et al., 2018)

3. **D*** (Dynamic A*) builds on A* by allowing the path to be dynamically adjusted as the environment changes. This feature is especially important in real-world urban environments, where new obstacles such as pedestrians, delivery trucks, or other moving robots may appear unexpectedly. D* helps maintain an optimal and feasible path even under changing conditions by dynamically updating the path.

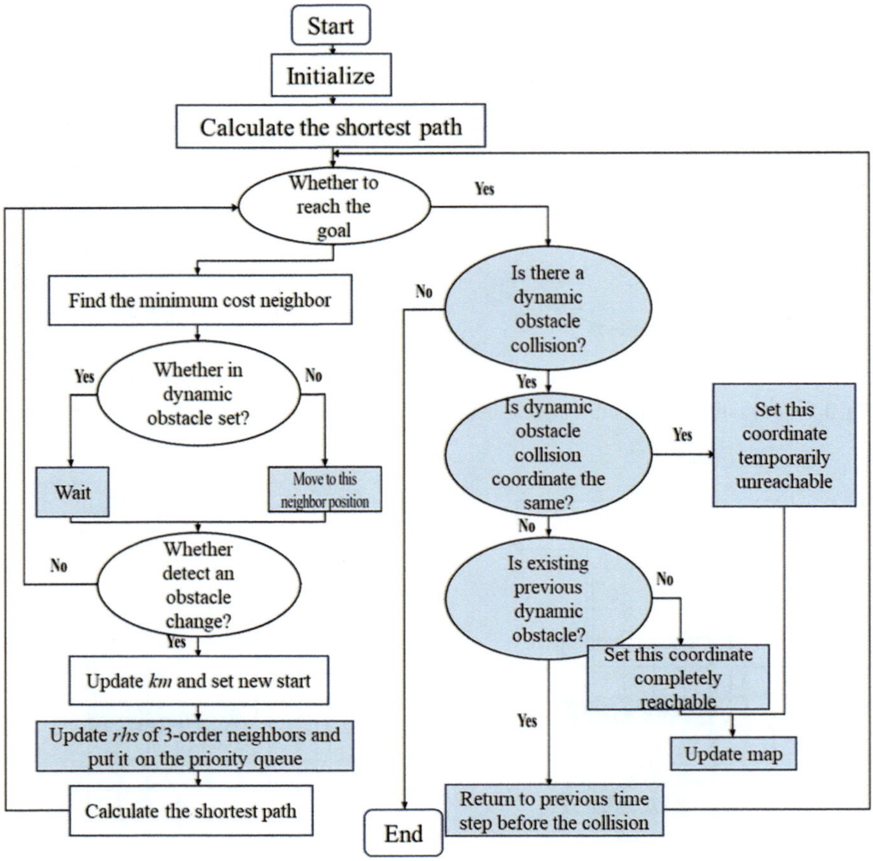

Fig. 3.3 D* algorithm diagram (*Source* Jin et al., 2023)

3.1.2 Comparison of Global and Local Route Planning

To effectively navigate complex environments, autonomous mobile robots rely on a combination of **global and local route planning**. Understanding these concepts is important for understanding the overall structure of autonomous driving systems.

- **Global route planning** is the process of finding an executable route from a starting point to a destination, taking into account the entire environment. This approach is suitable for high-level planning that is performed using existing maps. Global path planning allows robots to choose a path that allows them to get to their desired destination efficiently by avoiding major obstacles and choosing a path that is efficient in terms of distance and energy.
- **Local path planning**, on the other hand, focuses on avoiding immediate obstacles encountered during movement. It fine-tunes the path to account for dynamic changes in the environment, such as pedestrians suddenly entering the robot's path

or temporary obstacles appearing. Local path planning is often done by the robot continuously updating its understanding of the environment through real-time sensor data, including lidar and camera feeds.

For autonomous driving to work successfully in pedestrian-centric urban environments, there must be synergy between global and local planning. Global planning provides macro-level overview and guidance, while local planning provides micro-level adjustments to deal with unexpected changes.

3.1.3 Key Technology for Supporting Path Searching

Autonomous navigation capabilities rely heavily on a variety of technologies to facilitate path finding and environmental awareness, and Fig. 3.4 shows how LiDAR and cameras are used to generate data based on environmental awareness. Here we focus on two core technologies: **Graph Theory and Heuristic Methods**.

- **Graph theory** is the foundation of many navigation algorithms. In graph theory, an environment is represented as a set of interconnected nodes, where each node represents a location and each edge represents a path connecting two locations. The application of graph theory can effectively model an environment that can be navigated by algorithms such as A* and Dijkstra.
- **Heuristic methods** are used to improve the efficiency of search algorithms. A common heuristic is **Euclidean distance** or **Manhattan distance**, which is used to estimate the cost of reaching a goal. In the context of A* and D* algorithms, heuristics provide a way to make computations faster by significantly reducing the search space through informed guesses. This is particularly useful in real-time navigation scenarios where timely decision making is critical.

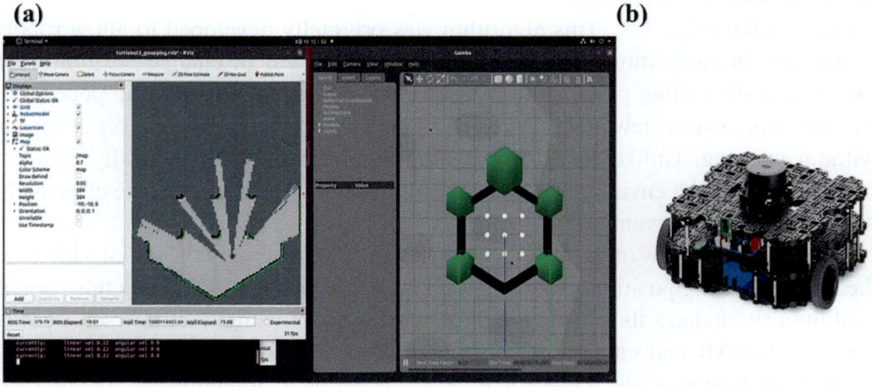

Fig. 3.4 **a** Turtlebot's LiDAR and camera-based data visualization, **b** Turtlebot

Building a reliable navigation system for autonomous mobile robots requires the integration of both existing algorithms and cutting-edge technologies. From the basic concepts of Dijkstra and A* algorithms to more advanced approaches such as D*, navigation systems have evolved to meet the complex requirements of pedestrian environments. As autonomous mobile robots become increasingly part of the urban fabric, the importance of efficient route planning that balances optimality, energy efficiency, and adaptability cannot be overstated. An effective combination of global and local route planning, supported by real-time data and intelligent algorithms, forms the foundation of future autonomous systems that aim to coexist harmoniously with pedestrians in vibrant urban landscapes.

3.2 Continuous Development of D* Algorithm Under Autonomous Mobile Robots

Autonomous mobile robots operating in pedestrian environments face numerous challenges that require advanced navigation strategies to adapt to dynamic and unpredictable conditions. This section provides an in-depth look at the latest research trends, development challenges, and key navigation strategies used to ensure efficient and reliable movement of autonomous mobile robots in these environments. In particular, this section focuses on the D* (Dynamic A*) algorithm and its modifications to improve efficiency, as well as its application in urban spaces with many pedestrians.

3.2.1 D Algorithm and Application*

D* (Dynamic A*) algorithm is one of the most important tools in the autonomous driving toolkit (Fig. 3.5). This algorithm was originally developed to allow robots to navigate in environments where obstacles may appear or change position after the initial path-finding process. This feature is particularly important in pedestrian environments where new obstacles, such as people or moving objects, often appear without warning. Unlike static path-finding algorithms, D* dynamically recalculates the path as the environment changes, allowing the robot to navigate effectively without manual intervention.

The basic workflow of the D* algorithm begins by calculating the initial path from the robot's starting position to the target point. As the robot moves along this path, it continuously updates its knowledge of the environment using real-time sensor data from the LiDAR and camera. When a new obstacle is detected, D* dynamically re-plans the path to adjust it to avoid the obstacle while maintaining an efficient path to the target. This feature makes D* suitable for applications where the environment is highly variable and requires rapid adaptation.

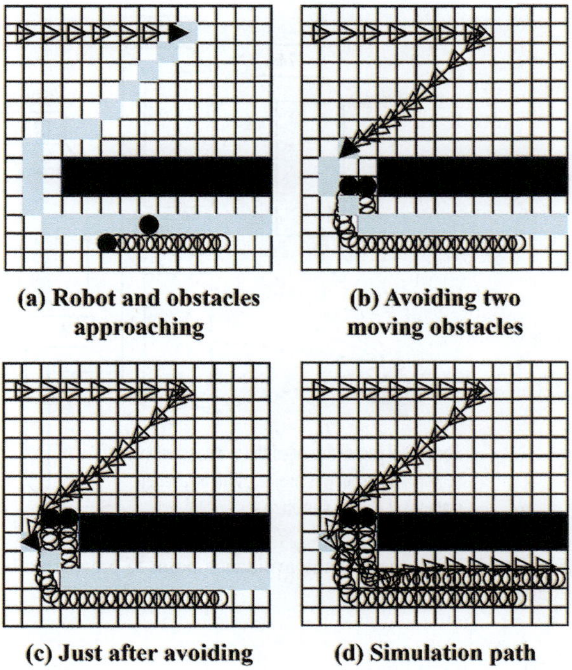

Fig. 3.5 Avoidance simulation of two obstacles (*Source* Jung et al., 2010)

3.2.2 Modifications to D for Pedestrian Interaction*

One of the main limitations of the existing D* algorithm is that it treats all obstacles the same, i.e. it tries to avoid all detected objects regardless of whether they could leave the path on their own. In real-world pedestrian environments, it has been observed that many people detect the approach of an autonomous mobile robot and take action to avoid it. Based on these observations, a modified version of the D* algorithm has been developed, as detailed in recent research.

The modified D* algorithm takes into account **human behavior patterns** when navigating pedestrian paths. The analysis of the Lidar data allows the determination of key parameters like **average distance and evasion angle**. These insights allowed us to modify the D* algorithm, which determines whether a robot should deviate from its planned path based on the likelihood that a pedestrian will naturally avoid the robot.

The result is an autonomous mobile robot that can maintain its original path when pedestrians are expected to move out of the way, and only recalculate a new path when necessary to optimize its movement. This modification reduces the frequency of unnecessary path deviations, improving both energy efficiency and overall motion speed. **Experimental test** results showed that the modified D* algorithm outperformed the existing method in terms of travel efficiency and time, demonstrating the importance of considering human–robot interaction in route planning.

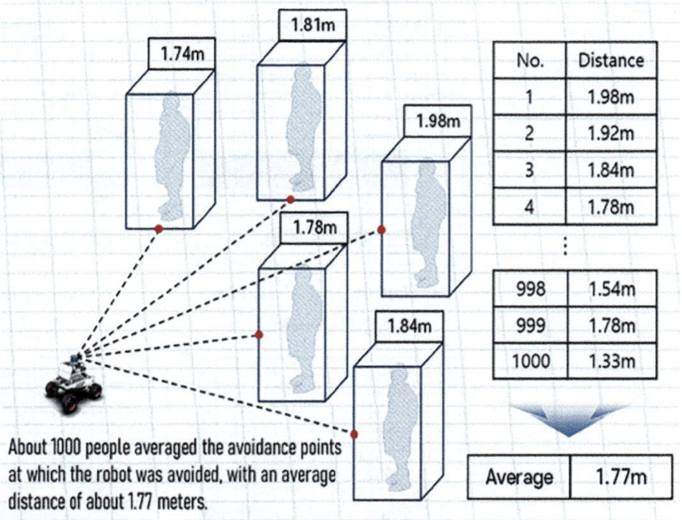

Fig. 3.6 Finding the average avoidance point to prevent collisions with pedestrians (Choi et al., 2024a)

In a related study, we improved the existing D* algorithm with the goal of optimizing the path of autonomous robots in pedestrian environments. The research focused on improving the efficiency of robot path planning by reflecting the behavioral patterns of pedestrian perception and avoidance of autonomous robots. To this end, LiDAR and camera sensors were used to collect and analyze pedestrian trajectory data to derive the average distance (1.77 m) and avoidance angle at which pedestrians start to avoid the robot (Fig. 3.6).

Based on this data, the study modified the D* algorithm so that it maintains the existing path when the pedestrian's intention to avoid is clear, and the robot calculates an avoidance path only when this is not the case. Figure 3.7 visually illustrates the working principle of the modified D* algorithm and explains the process of determining whether the robot should change its path based on the pedestrian's avoidance distance and angle.

The experimental results show that the modified D* algorithm reduces the total distance and time traveled by the robot compared to the original algorithm, while improving the average travel speed and energy efficiency. Notably, in about 76% of the interactions, the pedestrian avoided the robot first, demonstrating that algorithmic design that takes human behavior into account can contribute to reduced energy consumption and improved locomotion performance. This approach is a path optimization method that enables autonomous robots to operate efficiently and safely in pedestrian environments, with important implications for increasing the practicality and commercial viability of autonomous mobile robots.

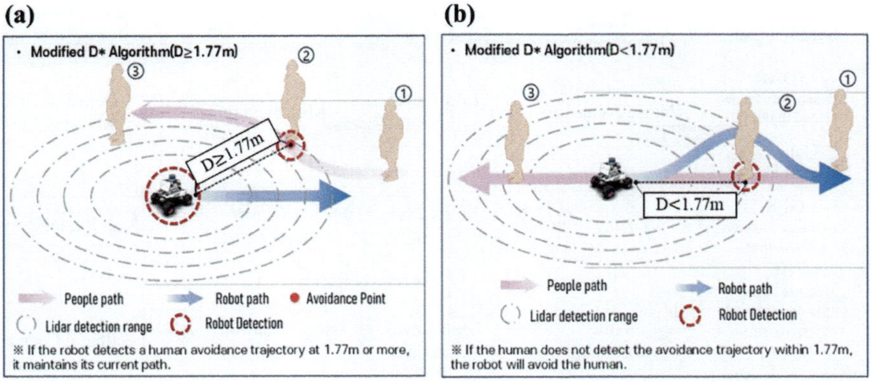

Fig. 3.7 Modified D* algorithm driving behavior: **a** D ≥ 1.77 m, **b** D < 1.77 m (Choi et al., 2024a)

3.2.3 Challenges and Future Directions

Despite progress in modifying navigation algorithms such as D* and using reinforcement learning for energy optimization, there are still significant challenges to ensuring that autonomous mobile robots are robust enough for real-world deployment. The **unpredictability of pedestrians**, variable weather conditions, and limitations of sensor technology all complicate navigation.

In future research, advances in sensor fusion technology are expected to play an important role in addressing these issues. **Sensor fusion** integrates input data from multiple sensors to provide a more accurate understanding of the environment, which can greatly improve the reliability and safety of robot navigation. For example, Fig. 3.8 shows an environment recognition procedure using a variety of sensors. It includes the entire process from data acquisition using LiDAR, cameras, GPS, IMU, and ultrasonic sensors, to data pre-processing, sensor fusion, environment modeling, and finally decision making and path planning.

Another notable development is the integration of sophisticated machine learning models and the use of edge computing technologies. Machine learning models can help predict pedestrian behavior more accurately, while **edge computing** can increase the efficiency of navigation strategies by reducing the latency of real-time data processing.

In conclusion, advanced navigation strategies such as the modified D* algorithm and the use of reinforcement learning for energy efficiency represent promising advances in the development of autonomous driving systems. By continuously improving these approaches and addressing current limitations, we can develop autonomous mobile robots that can be seamlessly integrated into pedestrian-centric urban environments, ultimately contributing to safer and more efficient urban mobility.

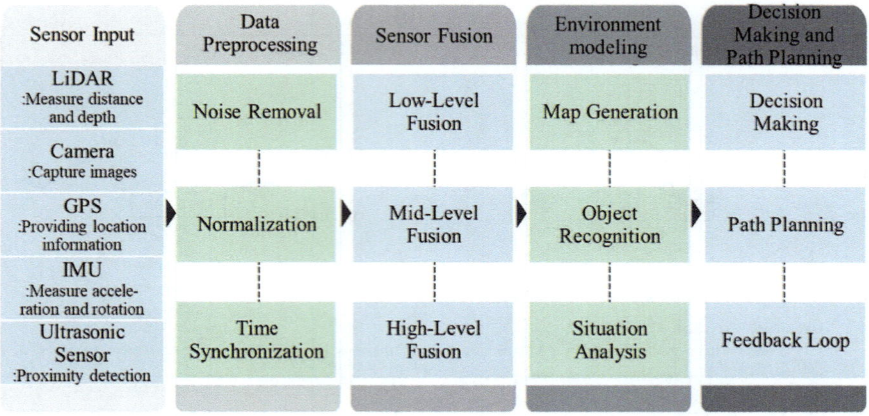

Fig. 3.8 The process of environmental recognition using multiple sensor

3.3 Real-Time Route Planning and Energy Efficiency Considerations

For autonomous mobile robots, real-time path planning is particularly important to ensure efficient and safe navigation, especially in complex urban environments with many pedestrians, cyclists, and other obstacles. To implement a robust real-time path planning system, autonomous mobile robots must not only make dynamic navigation decisions, but also consider energy efficiency, as they are often constrained in their operations due to limited battery capacity. This section focuses on the importance of real-time data processing, technologies that facilitate dynamic path adjustment, and strategies for improving energy efficiency in autonomous path planning.

3.3.1 Real-Time Route Planning in Dynamic Environments

Real-time route planning is a very important element for autonomous mobile robots, especially for robots operating in urban environments with a high concentration of pedestrians. Unlike static environments, these robots must be able to react immediately to unexpected changes, such as sudden obstacles or traffic light changes, as pedestrians enter the path. The key to effective real-time planning is the ability to **quickly process sensor data** and dynamically update the planned route according to new information.

To do this, autonomous mobile robots continuously collect data from their environment using a variety of sensors, including **LiDAR, cameras, and ultrasonic sensors**. This data is then processed in real time to identify and classify obstacles, detect human movements, and predict future events. Algorithms such as **dynamic A*(D*)** are often used to facilitate dynamic route changes based on incoming sensor data. This process allows the robot to adapt to changes in the environment without stopping its movement, contributing to smoother and safer navigation.

One of the main challenges of real-time path planning is managing the computational complexity associated with the continuous processing of sensor inputs. The robot must balance fast response time with computational efficiency to avoid overloading the on-board system. Advances in **edge computing and sensor fusion technology** play a key role in enabling autonomous mobile robots to better manage these computational demands (Kabir et al., 2024). **Edge computing** allows data to be processed closer to the sensor source, reducing latency and improving the responsiveness of navigation systems (Liu et al., 2019). On the other hand, **sensor fusion** combines data from multiple sources to create a more comprehensive and accurate model of the environment, improving the decision-making capabilities of robots (Yeong et al., 2021).

3.3.2 Energy Efficiency of Autonomous Mobile Robots

Energy efficiency is another important consideration in autonomous driving, especially for robots operating in environments where frequent recharging is impractical. Route planning decisions have a direct impact on a robot's energy consumption, as factors such as distance, speed, and number of maneuvers can all affect overall energy consumption. In pedestrian environments, where robots must frequently stop, change direction, or take detours, energy consumption can be significant.

To increase energy efficiency, it is important to develop an algorithm that can find the **most energy efficient route**, not just the shortest route. For example, algorithms such as **Dynamic A*** can be modified to favor smoother trajectories that minimize sharp turns and frequent stops to save energy. **Reinforcement learning methods** such as **Policy Optimization for Short-Distance (PPO)** have also shown promise in optimizing the energy efficiency of autonomous mobile robots by allowing them to learn optimal behavior based on previous experience. By applying reinforcement learning, robots can adjust their navigation strategies to reduce energy consumption, especially in scenarios where obstacles are dense or interactions with pedestrians are frequent.

A recent study explored how PPO can be used to develop adaptive strategies that allow autonomous mobile robots to optimize their movements in real time (Choi et al., 2024a, 2024b). The key idea was to use **environmental data**, such as obstacle density and pedestrian flow, to dynamically adjust the robot's path in a way that minimizes energy consumption. For example, in areas with high pedestrian traffic, the robot can save energy by choosing a path that requires fewer sudden stops and accelerations. Experimental results show that this adaptive approach significantly improves energy efficiency compared to conventional static path planning methods.

In a related study, researchers proposed a method to dynamically adjust the LiDAR sensing range to increase the energy efficiency of autonomous mobile robots (AMRs).

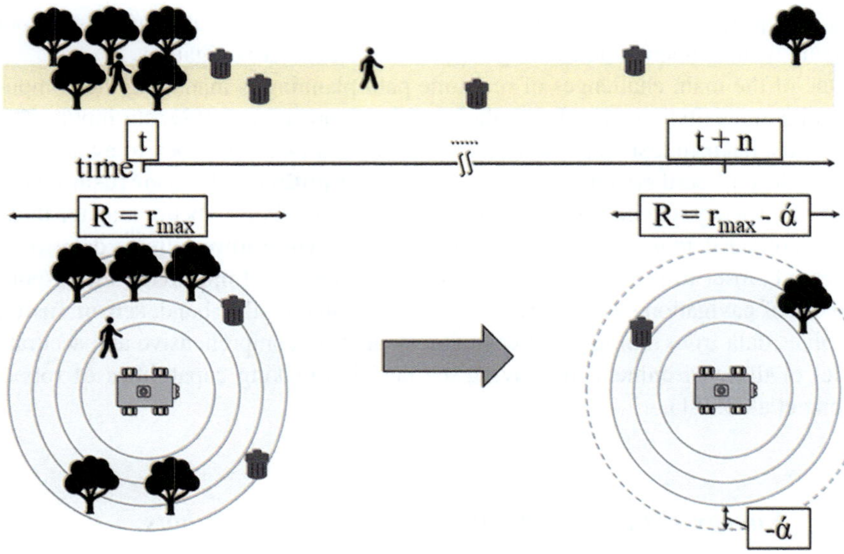

Fig. 3.9 Example of an energy-saving scenario by adjusting the LiDAR detection range (Choi et al., 2024b)

This study introduced the concept of "saturation" to optimize energy consumption by adjusting the LiDAR sensing range based on the density of obstacles.

Figure 3.9 shows the process of adjusting the LiDAR sensing range according to saturation. When the saturation is above 0.5, the maximum detection range is maintained to ensure safety, and when it is below 0.5, the detection range is reduced to reduce energy consumption. Figure 3.10 shows that as the saturation increases, the sensing range and energy consumption increase, and the additional travel time tends to decrease. For example, at a saturation of 0.05, the detection range was 120.52 m and the energy consumption was 15.8 Wh, but when the saturation increased to 0.45, the detection range increased to 376.29 m and the energy consumption increased to 19.64 Wh.

This study shows that LiDAR range adjustment can be used to efficiently manage the energy consumption of autonomous robots, achieving both safety and efficiency in urban environments.

3.3.3 Balancing Real-Time Responsiveness and Energy Efficiency

A fundamental challenge for autonomous driving is balancing **real-time responsiveness and energy efficiency**. Real-time route adjustments are essential for safety and navigation in dynamic environments, but these frequent adjustments often come at

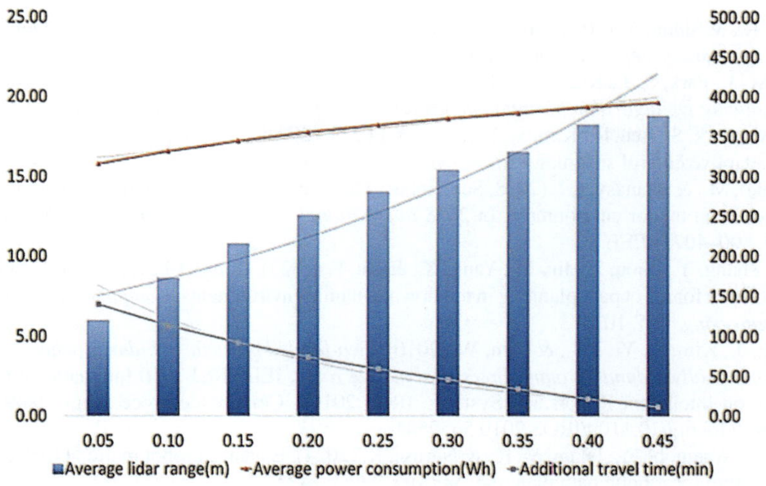

Fig. 3.10 Graph of Lidar range and power consumption analysis (Choi et al., 2024b)

the cost of increased energy consumption. Therefore, the goal is to develop navigation strategies that can effectively respond to environmental changes while minimizing unnecessary energy consumption.

One approach to solving this challenge is to use **predictive models** that anticipate environmental changes. For example, by analyzing the movement patterns of pedestrians, it is possible to predict when they will cross the road and adjust the robot's speed or trajectory in advance. This reduces the need for sudden route changes, resulting in smoother navigation and lower energy consumption. **Hierarchical path planning**, which divides path planning into multiple levels of decision making, can also help balance responsiveness and efficiency. At a high level, the robot plans an energy-efficient global path, while at a lower level it makes real-time adjustments to avoid obstacles without deviating significantly from the planned path.

In addition, **battery management systems (BMS)** play an important role in optimizing energy use by providing real-time feedback on the robot's energy status (Chellal et al., 2021). This feedback can be used to make path planning decisions, such as choosing a route with less energy consumption or reducing speed to save energy. Integrating BMS data with path planning algorithms enables robots to make informed decisions that extend their operating time while maintaining safety and efficiency (Chitre & Sinha, 2023).

References

Chellal, A. A., Lima, J., Gonçalves, J., & Megnafi, H. (2021, June). Battery management system for mobile robots based on an extended Kalman filter approach. In *2021 29th Mediterranean Conference on Control and Automation (MED)* (pp. 1131–1136). IEEE.

Chitre, R., & Sinha, A. (2023). *An enhanced RRT based algorithm for dynamic path planning and energy management of a mobile robot*. arXiv preprint arXiv:2310.20590

Choi, M. J., Park, S. J., Kim, S., & Lee, S. J. (2024a). *Human-aware trajectory optimization for enhancing D* algorithm for autonomous robot navigation* (May 2024). IEEE Access.

Choi, M., Park, S., Lee, R., Kim, S., Kwak, J., & Lee, S. (2024b). Energy efficient robot operations by adaptive control schemes. *Oxford Open Energy, 3*, oiae012.

Filipenko, M., & Afanasyev, I. (2018, September). Comparison of various slam systems for mobile robot in an indoor environment. In *2018 International Conference on Intelligent Systems (IS)* (pp. 400–407). IEEE.

Jin, J., Zhang, Y., Zhou, Z., Jin, M., Yang, X., & Hu, F. (2023). Conflict-based search with D* lite algorithm for robot path planning in unknown dynamic environments. *Computers and Electrical Engineering, 105*, 108473.

Jung, E.-J., Kim, S., Yi, B-J., & Kim, W. (2010). *Two-leveled obstacle avoidance scheme using a kinematically redundant omni-directional mobile robot*. IEEE/RSJ 2010 International Conference on Intelligent Robots and Systems, IROS 2010 – Conference Proceedings. 3949–3954. https://doi.org/10.1109/IROS.2010.5650584.

Kabir, R., Watanobe, Y., Islam, M. R., & Naruse, K. (2024). Enhanced robot motion block of a-star algorithm for robotic path planning. *Sensors, 24*(5), 1422.

Kim, D. H., Hai, N. T., & Joe, W. Y. (2018). A guide to selecting path planning algorithm for automated guided vehicle (AGV). In *AETA 2017-recent advances in electrical engineering and related sciences: Theory and application* (pp. 587–596). Springer International Publishing.

Liu, S., Liu, L., Tang, J., Yu, B., Wang, Y., & Shi, W. (2019). Edge computing for autonomous driving: Opportunities and challenges. *Proceedings of the IEEE, 107*(8), 1697–1716.

Yeong, D. J., Velasco-Hernandez, G., Barry, J., & Walsh, J. (2021). Sensor and sensor fusion technology in autonomous vehicles: A review. *Sensors, 21*(6), 2140.

Rehman, A. U., Awuah-Offei, K., Baker, D. A., & Bristow, D. (2019). Emergency evacuation guidance system for underground miners. In *SME Annual Meeting 2019* (pp. 19–100).

Chapter 4
Hardware Design and Scalability of Autonomous Robots

The successful integration of autonomous robots into urban environments depends on sophisticated hardware design and the scalability of these systems. In the ever-evolving landscape of modern cities, autonomous robots must not only perform their tasks efficiently but also seamlessly adapt to the physical infrastructure and social dynamics of urban spaces. Achieving this requires advanced hardware components that enhance functionality while also considering aesthetic and ergonomic design to improve user acceptance. Additionally, scalability is essential to ensure these robots can meet the diverse demands and complexities of various urban settings.

This chapter explores the modular and scalable design approach that is critical for contemporary urban applications. It examines the intricate process of developing hardware for autonomous robots, emphasizing a design philosophy that extends beyond simply assembling functional parts. The objective is to create robots that are not only safe and efficient but also visually appealing and adaptable to a wide range of environments and user needs. The discussion is structured into three primary sections, each offering a comprehensive perspective on key hardware design elements essential for deploying autonomous mobile robots in pedestrian-centric cities.

Section 4.1 focuses on the core hardware components and functional requirements of autonomous mobile robots. It provides an in-depth analysis of crucial elements such as sensors, actuators, power systems, and control mechanisms—components that collectively enable robots to operate smoothly in complex urban environments. Ensuring the seamless integration of these components is vital for both operational reliability and maintaining high safety standards in pedestrian-dense areas.

Section 4.2 highlights the balance between functionality, aesthetics, and ergonomics. For autonomous robots to be effectively integrated into urban spaces, they must be designed to be both accessible and user-friendly. Key design elements—including shape, color, and material selection—play a significant role in fostering public acceptance. Additionally, ergonomic considerations ensure that these robots can interact safely and comfortably with people in crowded city settings.

M. Choi and S. Lee, *Intelligent Moving Cities: Technological Leap and Social Integration of Autonomous Mobile Robots*,
https://doi.org/10.1007/978-981-96-8864-7_4

Section 4.3 explores the principles of modular design and scalability. A modular hardware architecture allows autonomous mobile robots to adapt to various roles and environments by enabling easy customization and maintenance. This flexibility is essential for deploying robots in multiple urban applications, including logistics, monitoring, and personal mobility. Furthermore, scalable designs ensure that these systems can evolve to accommodate shifting urban densities and dynamic operational demands.

Overall, this chapter provides a detailed examination of three fundamental aspects—core components, aesthetic and ergonomic considerations, and modular scalability—shedding light on both the challenges and opportunities in designing hardware for autonomous mobile robots. The ultimate goal is not only to meet functional requirements but also to enhance the urban landscape, contributing to safer, more efficient, and more livable cities.

4.1 Core Hardware Components and Functional Requirements

The design of autonomous mobile robots relies on the integration of various core hardware components that perform unique and complementary functions. These components must not only support basic functions of navigation, detection, and control, but also meet aesthetic and ergonomic requirements that facilitate human–robot interaction. This section examines essential hardware components, balance of functionality and aesthetics, and the importance of ergonomic design for smooth integration in pedestrian environments.

4.1.1 Essential Components Overview

Autonomous mobile robots provide optimal functionality through a powerful combination of sensors, actuators, power systems and controls. Figure 4.1 is an overview of these critical components and key roles.

Sensors: Sensors form the robot's "eye and ear" so that it can recognize the surrounding environment. This includes LiDAR, cameras, ultrasonic sensors, and RADAR, each providing unique insight into obstacles, distances, and object identification. LiDAR is especially useful for high-resolution mapping and obstacle detection, and the camera provides contextual information about the environment, such as visual signals for pedestrian activity and location determination (Maulana et al., 2018; Mochuard et al. 2023).

Actuator: The actuator is responsible for converting decisions from the robot's controller into physical motion. This includes a motor, servo, wheel, or joint that

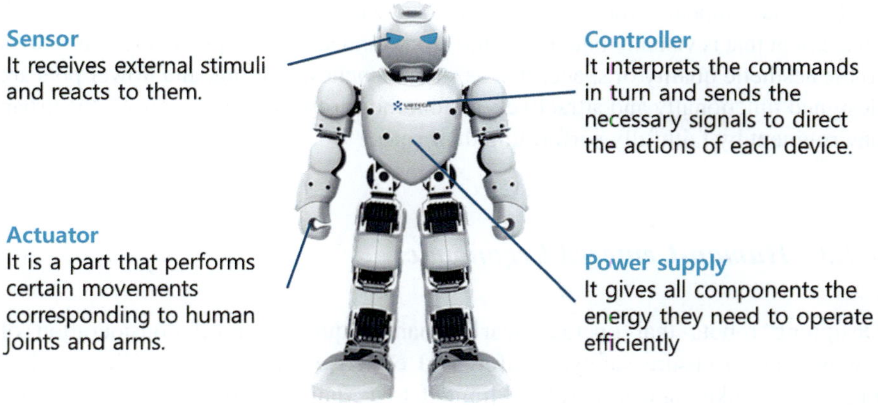

Fig. 4.1 The components of the robot

controls the movement, operation, and stability of the robot The actuator is included. In pedestrian environments, actuators must be precise to ensure smooth exploration and ability to dynamically respond to changing conditions (Pratt, 2002).

Power Supply: The power system is the lifeblood of an autonomous mobile robot that provides the energy required for all components to operate efficiently. Battery technology plays a key role, and lithium-ion batteries are most commonly used depending on energy density and rechargeability (McNulty et al., 2022). In particular, effective power management is important in order for robots to operate autonomously for a long time without frequent recharge in pedestrian-centered urban environments.

Controller: The control unit acts as the robot's brain, processes sensor inputs, and executes commands on the actuator. The controller is equipped with a CPU or GPU-like computing device that enables real-time data processing and decision-making. The control unit incorporates sophisticated algorithms for exploration, recognition, and planning to enable the robot to operate safely and effectively in complex environments.

4.1.2 Balance of Function and Aesthetic Sense

The key requirement for autonomous robots is functionality, but aesthetics also plays an important role, especially in the pedestrian environment in the city center. The robot's appearance can have a significant impact on the public's acceptance of the robot (Min et al., 2002). Robots designed with efficiency in mind are highly functional, but may lack the attractiveness to induce positive participation by pedestrians. Therefore, design must combine aesthetics and engineering to create a device that is not only efficient but also visually enjoyable and familiar.

The visual appeal of robots is also consistent with the concept of "urban furniture," an element that is visually integrated into urban landscapes and positively contributes to the aesthetic quality of space. Just as street benches, lighting, and flower pots are designed functionally and attractively, autonomous robots should improve the urban environment by faithfully performing their roles.

4.1.3 Human-Centered Ergonomics

Designing robots that operate near humans requires careful consideration of ergonomics to ensure safety, comfort and effective interaction, as illustrated in Fig. 4.2. Unlike industrial robots limited to certain environments, autonomous mobile robots must be designed to explore public places and safely interact with unpredictable pedestrian behaviour.

Proximity and Interaction: Hardware design must take into account human interaction at a fundamental level. This includes recognizing when the robot needs to slow down, stop, or change its route to maintain a comfortable distance from the pedestrian and prevent collisions and inconveniences. Proximity sensors and tactile interfaces play an important role in achieving these goals.

Ergonomic Interface: Robots designed for urban environments may include user interfaces for communication, such as touch screens and LED indicators. These

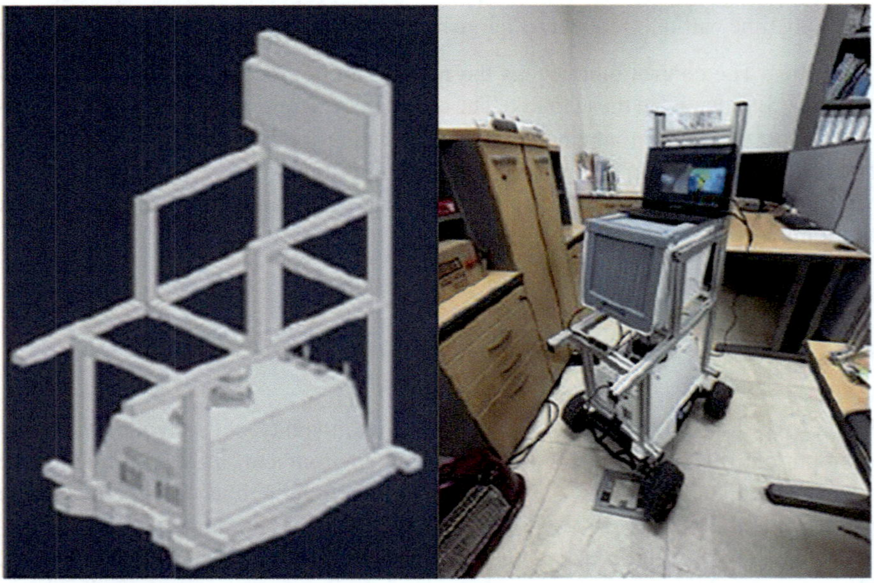

Fig. 4.2 Example of a robot that fits the user's characteristics: Porter robot

interfaces should be ergonomically arranged for easy access by people with various kidneys, including children and the disabled. Consideration of reach, viewing angle, and ease of use is an important aspect of designing ergonomic hardware interfaces.

Secure hardware structure: The physical structure of the robot must also prioritize pedestrian safety. Soft materials and protective down jackets can help reduce the risk of injury in case of accidental contact. Instead of minimizing sharp corners or protrusions that can cause injury, the design should focus on round, soft contours that can be safely contacted in crowded environments.

Motion Characteristics: Mechanics of robot motion, such as acceleration and deceleration, must also be considered from an ergonomic point of view (Shin et al., 2008). Sudden or sudden movements can surprise pedestrians or cause discomfort. Therefore, checking whether robot movements are soft and predictable is a key factor for humans and robots to coexist in harmony in urban space.

In conclusion, designing autonomous mobile robot hardware requires a complex balance of core functions, aesthetic attractiveness, and ergonomic considerations as well as simple technology. By integrating these elements, autonomous robots can be successfully integrated into pedestrian-centered urban environments to improve the quality of urban life and contribute to the broader goal of creating a sustainable and human-friendly urban space.

4.2 Modular Design and Customized Scalability

The concept of modular design is essential to creating adaptive, maintenance, and autonomous mobile robots that can play various roles in urban environments. Modular hardware not only facilitates scalability, but also enables the customization required to meet a variety of consumer needs and contextual requirements. In this chapter, we look at the advantages of modularity, the implications for user adaptability, and the role of modularity in expanding autonomous mobile robots to suit various urban dynamics.

4.2.1 Modular Hardware Architecture

Modular hardware architecture involves designing components in a way that is easy to assemble, upgrade, or replace without affecting the rest of the system. This design philosophy offers significant benefits for the distribution and maintenance of autonomous mobile robots, especially in complex and changing urban environments, as shown in Fig. 4.3.

A similar modular approach can be seen in Kia's PBV concept, depicted in Fig. 4.4.

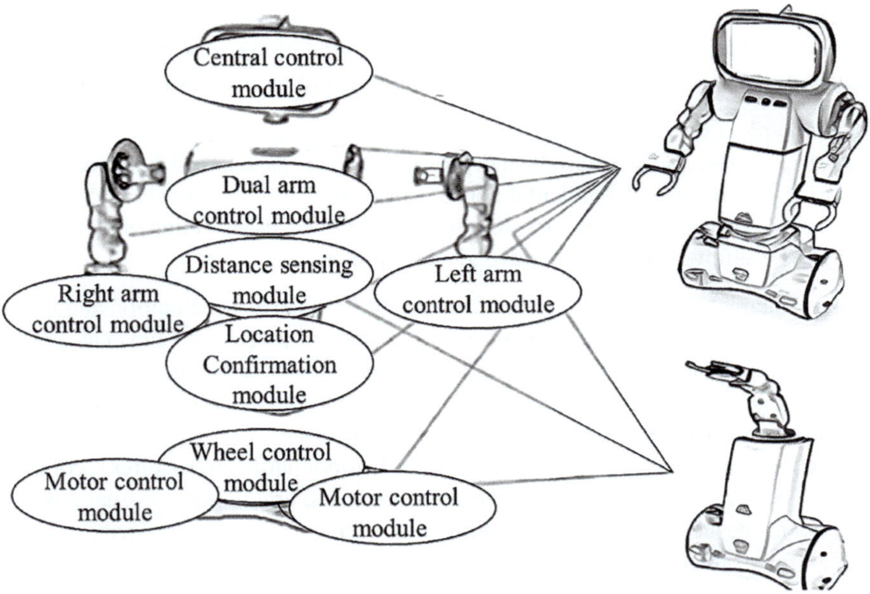

Fig. 4.3 Modular hardware (*Source* Korean Agency for Technology and Standards)

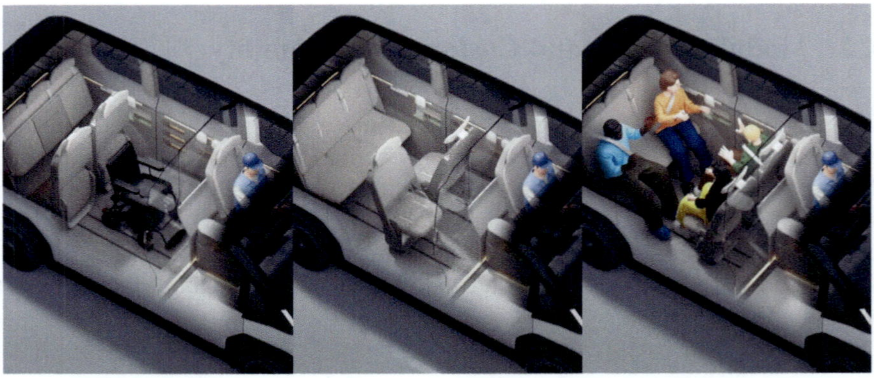

Fig. 4.4 Example of using a module in a PBV (*Source* Hyundai Motor Group)

- **Ease of maintenance and upgrade**: Modular approach for sensors, batteries Allows you to independently service or upgrade key components such as systems, actuators, etc. This significantly reduces robot downtime and can be updated efficiently as technology advances. For example, upgrading a camera module to a high-resolution version or replacing a damaged motor can be done quickly without replacing the entire robot.
- **Flexibility for a variety of applications**: Different urban environments and applications also require different functions for autonomous mobile robots. Modular

robots can easily replace modules and adjust them to specific tasks such as monitoring, cleaning, logistics, and support. This adaptability is consistent with the vision of deploying a multifunctional robot in the city that can handle everything from last mile delivery to personal mobility support.

- **PBV (Purpose Based Vehicle) Concept Example**: Modularity Concepts can be found in Kia Motors' Purpose Built Vehicle (PBV) designed for use in a variety of other applications besides traditional driving. The PBV model can transform the underlying vehicle structure to suit certain functions, such as freight transport or passenger movement, as required by the user. Similarly, modular autonomous robots can be adjusted to meet the dynamic requirements of urban environments, effectively broadening the potential range of use.

4.2.2 Adaptability to User Needs

In rapidly evolving urban environments, the ability to customize robot hardware to meet the needs of specific users is critical. Modular design plays an important role in providing the flexibility and adaptability needed for a variety of applications beyond traditional urban navigation.

- **Customized modules tailored to your specific needs**: Modular hardware allows you to easily install and replace feature modules to change the use of the same robot for a variety of applications. For example, a logistics robot can be equipped with a cargo handling module to transport goods during the day and switch to a monitoring module for night patrol. This flexibility maximizes the utilization of autonomous robots and provides cost-effective and versatile robots for urban operators and companies (Catalan et al., 2021).
- **Improved consumer interaction**: Modularity improves consumer interaction and satisfaction by allowing robots to be customized to meet the specific needs of various end users. For example, a personal auxiliary robot may include modules that can be easily adjusted to suit your personal preferences, such as a variety of payload capacities, interactive screens, or other auxiliary features.
- **Urban robotics for the future**: The rapid advancement of technology has made it difficult to keep robot solutions up to date. Modular hardware enables progressive improvements, enabling the integration of the latest advanced components without a complete overhaul. Thanks to these future-oriented features, urban robot infrastructure can be developed organically while responding to new challenges without huge replacement costs.

4.2.3 Scalability for Urban Dynamics

Scalability is an important consideration when deploying autonomous mobile robots throughout the city. Modular design provides the foundation for robots to scale

Fig. 4.5 Low density suburbs and high density city center

across different types of urban landscapes, from low-density suburbs to high-density metropolitan areas. Scalability is a key consideration when deploying robots across diverse urban landscapes, as illustrated in Fig. 4.5.

- **Hardware expansion for different environments**: A modular approach allows you to scale the size and capacity of autonomous robots to meet the population density and functional demands of specific urban areas. For example, in high-density areas, small robots with enhanced sensor arrays are preferred for safe exploration, while in more open environments, large robots with a wider range and larger payload capacity may be more effective.
- **Configurable sensor sets**: Modularity also applies to sensor configurations, allowing the robot to install a variety of sensor sets according to operational requirements (Kümmerle et al., 2015). For example, additional cameras and proximity sensors can be installed in robots in areas with high pedestrians to increase recognition and safety, while long-distance LiDAR can be focused on mapping and navigation in isolated areas.
- **Network for urban integration system**: Scalable modular hardware design enables collaborative urban solutions Supports the concept of robot systems. Smaller modules can work together by configuring interconnected vehicles to clean public places, provide information, and efficiently distribute deliveries. This scalability extends beyond individual robots to a network of robots that work together to effectively serve urban environments.

In conclusion, modularity and scalability are the cornerstones of autonomous mobile robot design that can adapt to the diverse and changing needs of urban environments. Modular hardware design supports simple maintenance, customization, and ability to scale or scale back to meet contextual requirements, helping these robots integrate seamlessly into urban structures and ultimately play a central role in improving robot usefulness and public acceptability.

4.3 Durability and Safety Design Considerations

Ensuring durability and safety is a very important factor in designing autonomous mobile robots, especially for robots that want to operate in urban environments where human interaction occurs all the time. This section discusses key design considerations to improve both hardware robustness and safety for pedestrians and other city dwellers. Durability and safety are not only important for efficient functions, but also essential to win public trust and ensure the wide adoption of autonomous robots in everyday urban environments.

4.3.1 Robustness in a Variety of Environments

Urban environments pose a variety of challenges for autonomous mobile robots, ranging from unpredictable weather conditions to unequal surfaces and various operating settings. Hardware must be robust enough to handle these issues reliably to ensure consistent performance, as shown in Fig. 4.6.

- **Weather resistance**: Robots operating in urban environments must withstand a variety of weather conditions, including rain, snow, heat and humidity. This requires careful selection of materials and protective design elements. For example, waterproof seals and weatherproof coatings are essential to protect sensitive electronic and mechanical components. In addition, robots must test under extreme conditions to verify reliability.
- **Shock absorption and shock resistance**: Urban environments can be difficult to predict due to the high number of obstacles and the possibility of accidental collisions. Designing hardware that includes shock absorption materials and components or that can absorb shock can reduce damage caused by minor shocks and ensure continuous operation without frequent maintenance (Seo et al., 2019). For example, rubber bumpers and flexible external cases can absorb shocks and reduce the risk of internal damage.

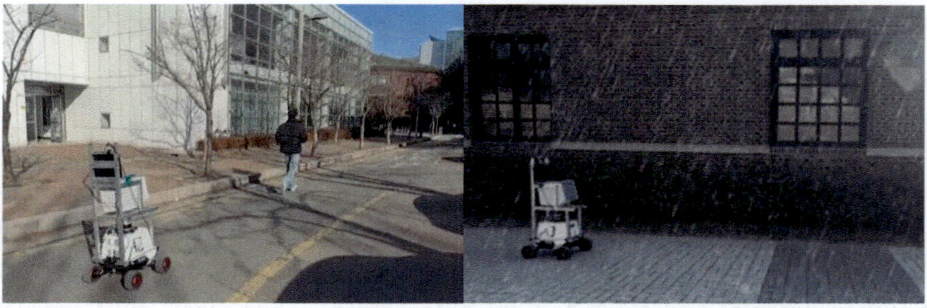

Fig. 4.6 Autonomous robots in various conditions

- **Material durability**: Materials used to build autonomous mobile robots must be light and durable. Composites such as carbon fibers and high-grade polymers are often used because of their high weight specific strength. Durability extends to abrasion resistance, and parts such as wheels and trucks must be resistant to abrasion due to persistent use on asphalt, gravel roads, or other urban surfaces.

4.3.2 Safety Features for Interaction with People

Safety is the most important thing when autonomous mobile robots operate near people. Safety design includes not only the robot's physical structure but also its ability to interact appropriately and predictably with the environment.

- **Proximity sensors and collision prevention**: Autonomous mobile robots must be equipped with advanced proximity sensors and collision prevention systems that can detect obstacles, including pedestrians, in real time. LiDAR, ultrasonic sensors, and cameras are very important for 360° recognition of the surrounding environment. These sensors should work with the robot's navigation algorithm to ensure pedestrian safety by reacting immediately to avoid collisions (Choi et al., 2024).
- **Soft look and human-friendly design**: The robot's physical design should use soft materials and round corners to minimize the risk of injury during accidental human contact. Padded covers and rubberized appearances can make the robot pedestrian-friendly and reduce the risk of injury caused by accidental contact with the robot (Park et al., 2011). These design considerations are especially important in crowded urban areas where space is limited and there is a high possibility of interaction.
- **Hearing and visual signals**: To communicate effectively with pedestrians, autonomous robots must include both auditory and visual signals representing intentions. Lights such as LED lights can be used to signal a change of direction or stop, and soft auditory signals can signal the presence of robots without disturbing pedestrians. These signals must be intuitively and easily understood by all age groups to prevent confusion and ensure safe coexistence.

4.3.3 Testing and Certification Standards

Strict testing and compliance with authentication protocols are essential for autonomous mobile robots to meet the highest levels of durability and safety standards. Compliance with these standards not only ensures stable performance, but also increases public confidence in technology.

- **Environmental and stress testing**: Autonomous mobile robots must conduct extensive environmental and stress testing to assess their ability to withstand

Fig. 4.7 Examples of field testing sites (*Source* UCL PEARL Center)

a variety of urban conditions. This includes testing for extreme temperature, humidity, dust, and vibration. Stress tests check if the robot can withstand the physical needs of continuous movement and interaction without frequent failures in urban environments.

- **Safety Certification**: Relevant safety certification and compliance with standards are essential for deploying autonomous mobile robots in public places. This may include ISO standards and electronic stability and electromagnetic compatibility certification related to robot safety. Once these certifications are met, you can ensure that the robot is designed and tested to operate safely around people.
- **Field Test and Public Demonstration**: Conducting on-site tests and public demonstrations prior to full-scale deployment is an important step in verifying the durability and safety of the robot. These tests help identify potential weaknesses in real-world environments that may not appear in laboratory tests. Public demonstrations also help improve acceptability and reliability by making communities familiar with robots, as demonstrated in Fig. 4.7.

In conclusion, designing an autonomous mobile robot with durability and safety requires solving unique problems in the urban environment and ensuring that the robot operates stably and safely around people. Robust materials, shock absorption, human-friendly design capabilities, and strict test standards all contribute to adapting these robots to urban deployments. By prioritizing durability and safety, designers can help autonomous mobile robots become a reliable and effective part of the urban landscape.

References

Catalan, J. M., Blanco, A., Bertomeu-Motos, A., Garcia-Perez, J. V., Almonacid, M., Puerto, R., & Garcia-Aracil, N. (2021). A modular mobile robotic platform to assist people with different degrees of disability. *Applied Sciences, 11*, 7130. https://doi.org/10.3390/app11157130

Choi, M. J., Park, S. J., Kim, S., & Lee, S. J. (2024). Human-aware trajectory optimization for enhancing D* algorithm for autonomous robot navigation. *IEEE Access*. https://doi.org/10.1109/ACCESS.2024.3430352

Kümmerle, R., Ruhnke, M., Steder, B., Stachniss, C., & Burgard, W. (2015). Autonomous robot navigation in highly populated pedestrian zones. *J Field Robot, 32*, 565–589. https://doi.org/10.1002/rob.21534

Maulana, I., Rusdinar, A., & Priramadhi, R. A. (2018). Lidar application for mapping and robot navigation on closed environment. *Journal of Measurements, Electronics, Communications, and Systems, 4*, 20–26. https://doi.org/10.25124/jmecs.v4i1.1696

McNulty, D., Hennessy, A., Li, M., Armstrong, E., & Ryan, K. M. (2022). A review of Li-ion batteries for autonomous mobile robots: Perspectives and outlook for the future. *Journal of Power Sources, 545*, 231943. https://doi.org/10.1016/j.jpowsour.2022.231943

Min, J. W., Lee, K., Lim, S. C., & Kwon, D. S. (2002). Human-friendly interfaces of wheelchair robotic system for handicapped persons. *IEEE/RSJ International Conference on Intelligent Robots and Systems, 2*, 1505–1510. https://doi.org/10.1109/IRDS.2002.1043968

Mochurad, L., Hladun, Y., & Tkachenko, R. (2023). An obstacle-finding approach for autonomous mobile robots using 2D LiDAR data. *Big Data and Cognitive Computing, 7*, 43. https://doi.org/10.3390/bdcc7010043

Park, J. J., Haddadin, S., Song, J. B., & Albu-Schäffer, A. (2011). Designing optimally safe robot surface properties for minimizing the stress characteristics of human-robot collisions. *International Conference on Robotics and Automation*, 5413–5420. https://doi.org/10.1109/ICRA.2011.5980282

Pratt, G. A. (2002). Low impedance walking robots. *Integrative and Comparative Biology, 42*, 174–181. https://doi.org/10.1093/icb/42.1.174

Seo, Y. S., Cho, S. J., Lee, J. Y., Park, C., Kim, U., Lee, S., & Song, S. H. (2019). Human-mimetic soft robot joint for shock absorption through joint dislocation. *Bioinspiration & Biomimetics, 15*, 016001. https://doi.org/10.1088/1748-3190/ab46f7

Shin, D., Sardellitti, I., & Khatib, O. (2008) A hybrid actuation approach for human-friendly robot design. *International Conference on Robotics and Automation*, 1747–1752. https://doi.org/10.1109/ROBOT.2008.4543453

Chapter 5
Various Applications of Autonomous Robots

With the rapid advancement of autonomous robot technology, urban environments are undergoing significant transformations, and innovative solutions are emerging to meet the evolving needs of cities and industries. Once considered a concept of the future, autonomous robots are now a reality, playing crucial roles in public safety, personal mobility, and commercial and industrial sectors. These robots can autonomously navigate complex environments, respond to dynamic situations, and interact seamlessly with both people and infrastructure.

The growing demand for automation in urban environments is driven by multiple factors, including the need to improve public safety, reduce labor-intensive tasks, enhance efficiency, and provide more inclusive mobility solutions. As urban populations increase and pressure on public services and infrastructure intensifies, there is a growing need for technologies that can support and augment existing systems. Autonomous robots, with their ability to operate independently and adapt to changing environments, are well-positioned to address these challenges. By integrating autonomous robots into various aspects of urban life, cities can become more resilient, more responsive, and ultimately offer a higher quality of life to their residents.

This chapter explores the diverse applications of autonomous robots across multiple domains, emphasizing their role in creating safer, more accessible, and more efficient urban environments. By examining their deployment in public safety and security, personal and public mobility solutions, and commercial and industrial sectors, this chapter highlights the transformative potential of these technologies. Autonomous robots are not confined to specific tasks; their adaptability allows them to provide valuable services across different areas of daily life, contributing to the development of smarter and more resilient cities.

Section 5.1 of this chapter focuses on the use of autonomous robots in public safety. These robots serve as highly effective tools in law enforcement, disaster management, and urban maintenance, performing functions such as surveillance,

M. Choi and S. Lee, *Intelligent Moving Cities: Technological Leap and Social Integration of Autonomous Mobile Robots*,
https://doi.org/10.1007/978-981-96-8864-7_5

emergency response, and infrastructure upkeep. By operating in hazardous or challenging environments, autonomous robots not only enhance safety but also allow human personnel to focus on more complex and strategic tasks.

Section 5.2 explores how autonomous robots are revolutionizing mobility solutions for both individuals and the public. Personal mobility robots provide a new level of independence for individuals with mobility challenges, helping them navigate urban spaces more easily. Additionally, autonomous robots are integrated into public transportation systems, bridging gaps between different modes of transit and offering first-mile and last-mile connectivity. This integration plays a crucial role in reducing reliance on private vehicles, alleviating traffic congestion, and promoting sustainable urban mobility.

Section 5.3 examines the impact of autonomous robots in commercial and industrial sectors, where they are significantly enhancing efficiency, cost-effectiveness, and safety. From logistics and delivery to agriculture and manufacturing, autonomous robots automate labor-intensive processes, minimize human error, and improve overall productivity. In warehouses, for example, robots optimize inventory management, while in agriculture, autonomous machines implement precision farming techniques to maximize crop yields and minimize resource consumption. These applications not only demonstrate the flexibility and versatility of autonomous robots but also underscore their potential to transform traditional industries and drive economic growth.

Overall, the applications of autonomous robots discussed in this chapter illustrate their potential to address some of the most pressing challenges modern cities face. By enhancing public safety, introducing innovative mobility solutions, and improving industrial efficiency, autonomous robots are becoming essential components of future urban environments. As these technologies continue to evolve, their role in building smarter, more sustainable, and more inclusive cities will become increasingly significant. This chapter provides a comprehensive overview of how autonomous robots contribute to these transformations, offering insights into both current implementations and future possibilities.

5.1 Utilization of Robots for Public Safety and Security

Autonomous robots have the potential to revolutionize public safety and security by providing consistent and efficient real-time support in urban environments. As cities grow and the demand for public safety increases, autonomous robots offer scalable solutions for maintaining order, supporting emergency response, and enhancing the quality of urban life. This section explores various ways autonomous robots are utilized for public safety and security, focusing on patrol and surveillance, disaster response, and cleaning and maintenance. These applications demonstrate how autonomous robots have transitioned from experimental technologies to essential components of urban infrastructure, showcasing their versatility in addressing diverse challenges.

5.1.1 Patrol and Surveillance Robots

One of the most prominent applications of autonomous robots in public safety is patrol and surveillance operations. Patrol robots are used to continuously monitor surrounding environments and identify potential threats through repeated observations (Basilico, 2022). These robots are increasingly deployed in various urban environments, such as parks, public squares, and residential areas, to enhance safety by providing real-time monitoring and response capabilities.

Equipped with various sensors, such as cameras and LiDAR, these robots effectively monitor their surroundings and detect abnormal activities (Andreasson et al., 2007; Mahalakshmi et al., 2023).

Autonomous patrol robots are particularly useful for complementing human security personnel by continuously surveilling large areas without fatigue. For example, in locations like shopping malls or airports, autonomous patrol robots navigate independently, monitor crowds, and detect suspicious behavior, alerting operators when intervention is required.

Furthermore, these robots leverage artificial intelligence to recognize patterns, analyze behaviors, and predict potential security threats based on accumulated data. Such predictive capabilities play a crucial role in enhancing proactive responses, preventing incidents before they escalate. Patrol robots are particularly valuable for nighttime surveillance, either replacing or complementing human personnel to reduce risks. In several East Asian cities, patrol robots have been deployed to enhance safety in public parks and residential areas during the night. These robots are also designed to operate effectively in hazardous environments such as narrow pathways or radiation zones (Kumar et al., 2024).

A notable example is the K5 autonomous security robot operated by Knightscope in the United States (Fig. 5.1). This robot is deployed in parking lots, corporate campuses, and public spaces, providing 24-h surveillance throughout the year to support human security teams. Capable of autonomous navigation and video capture, the K5 robot not only deters criminal activities but also supplies critical evidence required for investigations.

5.1.2 Disaster Response and Rescue Robots

Rapid response during disasters is critical for saving lives and minimizing damage. Autonomous robots have emerged as invaluable assets in disaster response and rescue operations, particularly in situations that are too dangerous or inaccessible for humans (SeungSub et al., 2017). These robots are used to assess disaster sites, search for survivors, and provide real-time data to emergency responders (Bogue, 2019).

One of the key advantages of autonomous robots in disaster response is their ability to navigate hazardous environments. Social Safety Robots and Life Search Robots are equipped with robust mobility systems that enable them to traverse debris,

Fig. 5.1 Knightscope robot for patrol and surveillance (*Source* https://www.knightscope.com/)

flooded areas, and other challenging terrains (Figs. 5.2 and 5.3). For example, the BEAR robot and Hibiscus robot (Fig. 5.2), have been deployed in scenarios such as building collapse sites and areas with hazardous chemical spills. These robots effectively overcome obstacles, quickly assess their surroundings, and support rescue operations while ensuring human safety. Additionally, autonomous robots equipped with thermal cameras can detect body heat, making them highly effective in locating survivors trapped under debris.

Autonomous robots are also gaining attention for their applications in firefighting and hazardous material handling. Firefighting robots, such as the RS1 model (Fig. 5.2), are designed to withstand extreme heat and navigate smoke-filled environments to assist firefighters. These robots can be remotely controlled to enter burning buildings, spray water, and suppress fires, all while keeping human firefighters out of harm's way. Furthermore, robots equipped with chemical sensors are used to detect and contain hazardous substances, reducing risks for human responders.

Autonomous robots also play a crucial role in post-disaster assessment and recovery efforts. After a disaster, these robots are utilized to map affected areas, evaluate structural integrity, and provide valuable data to support reconstruction efforts. They employ 3D mapping technology to create detailed models of damaged infrastructure, enabling engineers and planners to make more accurate decisions regarding repairs and rebuilding.

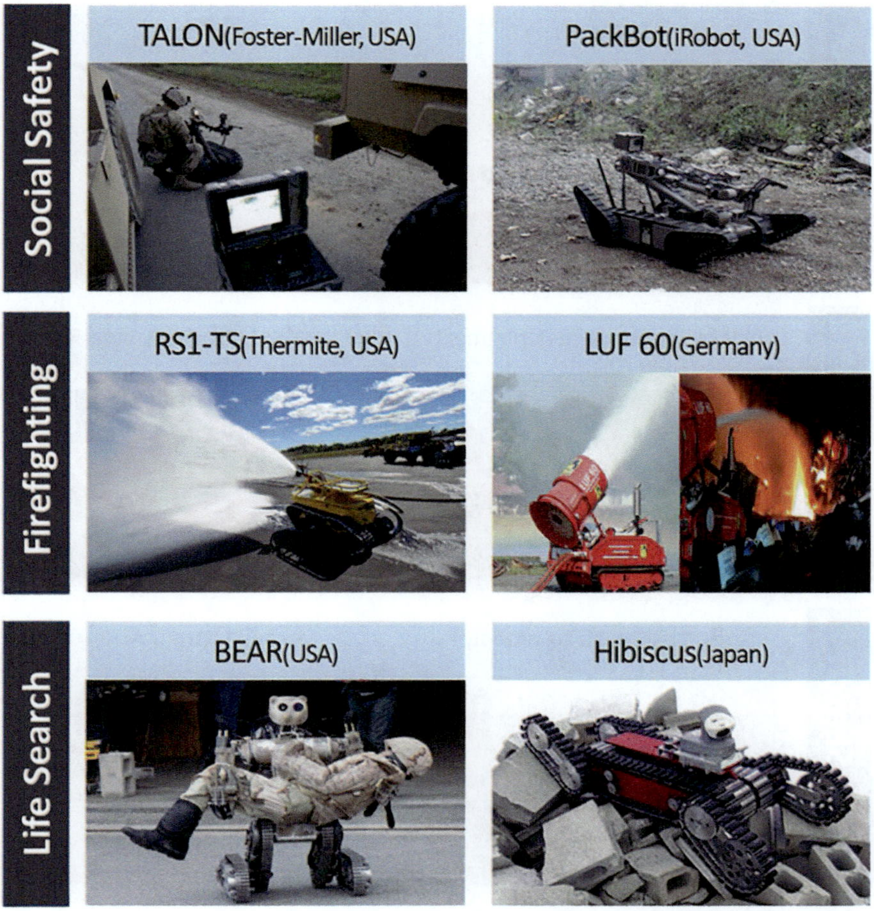

Fig. 5.2 Overseas disaster response robots (*Source* https://web.archive.org/web/20080808120417/http://www.fostermiller.com/lemming.htm) by TALON, PackBot (https://www.flirkorea.com/products/packbot/?vertical=ugs&segment=uis), RS1-TS (http://becauserobots.org/links/2017/05/robot-firefighter/), LUF 60 (https://www.chasesiam.com/products/b5-fire-fighting-vehicle/fire-fighting-robot/luf-60-fire-fighting-robot/), BEAR (https://www.army.mil/article/48456/robots_to_rescue_wounded_on_battlefield), Hibiscus (https://furo.org/en/works/hibiscus.html)

5.2 Cleaning and Maintenance Robots

Maintaining cleanliness and hygiene in public spaces is crucial for urban health and safety. Cleaning robots are now widely used in commercial settings such as offices, malls, airports, hotels, and hospitals (Memmesheimer et al., 2024). In large cities, where labor and maintenance costs are high, autonomous cleaning robots are increasingly deployed in parks, streets, transportation hubs, and public buildings for tasks like mopping, cleaning, and disinfection (Fig. 5.4).

Fig. 5.3 Domestic disaster response robots (*Source* https://www.inews24.com/view/115129) by DT3, Securo (https://www.clien.net/service/board/news/277399), LUF60 (https://www.incheon.go.kr/119/NE030401/2072520), Fire Control robot (https://www.hyundai.co.kr/news/CONT0000000000163672), RobHaz (https://m.post.naver.com/viewer/postView.naver?volumeNo=39911289&memberNo=44743697), ARMstrong (https://www.kaeri.re.kr/board/view?pageNum=1&rowCnt=10&no1=1301&linkId=11617&menuId=MENU00326&schType=0&schText=&boardStyle=Image&categoryId=&continent=&country=&schYear)

Autonomous cleaning robots gained prominence during the COVID-19 pandemic as tools to maintain hygiene and reduce virus spread. To limit human contact, robots were used for cleaning, with some countries using ultraviolet-based disinfection achieving 99.99% efficiency in hospital rooms within 15 min (Wang & Wang, 2021). Robots with disinfectant sprayers were deployed in high-contact areas like airports and hospitals, reducing infection risks for cleaning staff (Ruan et al., 2021).

Maintenance robots are also employed for infrastructure inspections, defect detection, and minor repairs. They are used to inspect bridges, tunnels, and other critical infrastructures, identifying cracks or corrosion with high-resolution cameras

Fig. 5.4 Cleaning robot platforms (*Source* https://www.directindustry.com/ja/prod/adlatus-robotics-gmbh/product-200716-2023748.html) by Adlatus CR700, VIGGO S100-N (https://korean.alibaba.com/product-detail/S100N-Autonomous-Robot-Sweeper-Cleaner-Intelligent-1601200323636.html)

and sensors. Regular automated inspections help detect issues early, reducing major failures and improving public safety.

In Singapore, autonomous cleaning robots are used to clean common areas in public housing complexes, such as corridors, staircases, and void decks. The cleaning robot LEOBOTS (Fig. 5.5), operates according to a predefined schedule and uses advanced sensors to navigate around obstacles, ensuring thorough cleaning without disrupting residents. By integrating autonomous cleaning solutions, cities can enhance the efficiency of public services while reducing the costs and labor associated with manual cleaning.

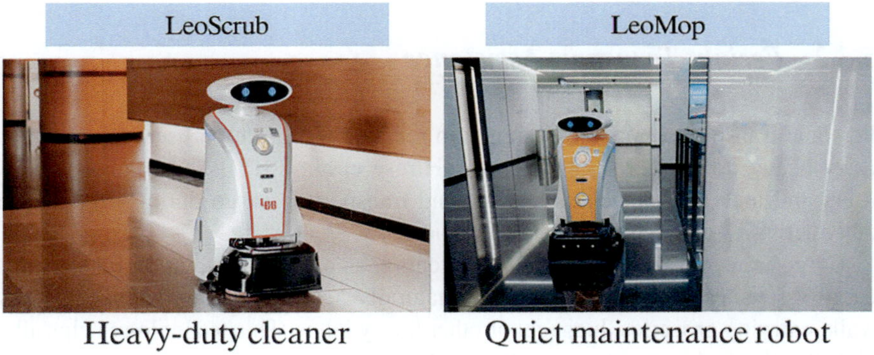

Fig. 5.5 Cleaning robots LEOBOTS (*Source* https://www.lionsbot.com/leobots/)

5.3 Personal and Public Mobility Solutions

Autonomous robots offer innovative solutions to enhance mobility for individuals and the public, with significant potential to facilitate the integration of personal transportation modes and public transit systems. These robots play a crucial role in creating more inclusive and accessible urban environments by providing mobility assistance, particularly for individuals with physical limitations, and enabling smoother access to public transportation.

5.3.1 Personal Mobility Robots

Personal mobility robots are designed to assist individuals who face challenges in using conventional transportation methods (Wang et al., 2012). Autonomous wheelchair that can navigate complex urban environments while avoiding obstacles and following the user (Fig. 5.6). Such robots are especially useful for the elderly, the disabled, or people who need help moving around.

Autonomous wheelchairs (Fig. 5.7), are integrated with advanced navigation systems, enabling them to understand urban structures and navigate seamlessly through various environments such as streets, parks, and public buildings. This enhances users' mobility and independence, allowing them to participate more actively in urban life. These robots go beyond providing mobility assistance, helping individuals with physical limitations to navigate urban environments more easily and effectively.

5.3.2 Freight Transport Assistance

Personal transport robots are designed to reduce the burden on users who need to carry heavy items over long distances (Hirose et al., 2015). These robots follow users and make tasks such as loading and transporting cargo during shopping or daily activities more convenient and efficient.

Cargo transport robots are gaining attention as a valuable solution for personal cargo assistance. Following robot that naturally accompanies individuals as they walk, carrying groceries, luggage, or other heavy items. This technology helps alleviate the physical strain of moving heavy objects and significantly enhances mobility, especially for the elderly or individuals with physical limitations (Fig. 5.8).

The following robot is designed to transport up to 10 kg of cargo while closely following its user. Equipped with advanced visual and sensory systems, this robot seamlessly navigates urban environments, maintaining proximity to the user and effectively avoiding obstacles. This technology can control the robot's actions

Fig. 5.6 Autonomous wheelchair

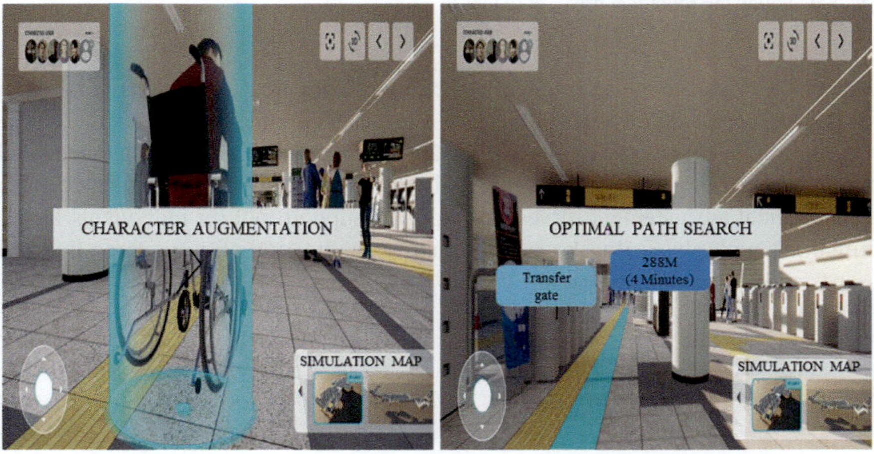

Fig. 5.7 Autonomous wheelchairs and advanced navigation systems for individuals with limited mobility

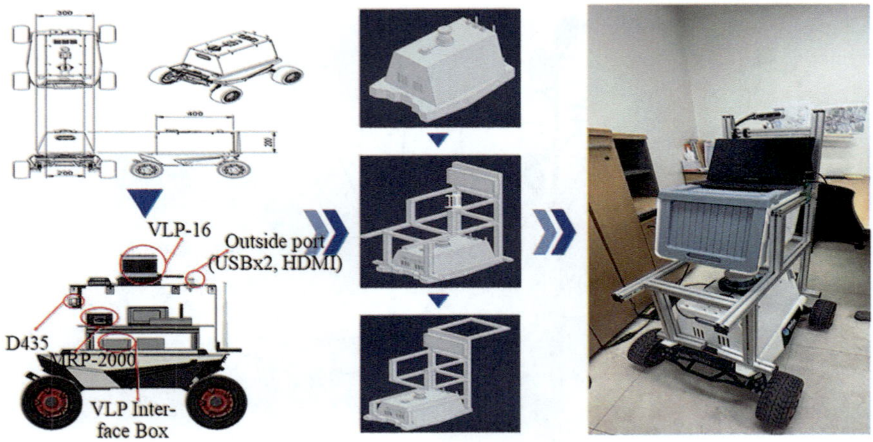

Fig. 5.8 Following robot design process

Fig. 5.9 Following robot that drives outdoors

through the finger's cognitive signals, and provides a practical and innovative solution for the personalized delivery of objects in various environments (Fig. 5.9).

It can move freely across diverse terrains such as sidewalks, roads, and speed bumps while trailing the user, significantly enhancing everyday convenience. This technological advancement serves as a valuable support tool for individuals facing difficulties with physical tasks, contributing to the creation of more accessible and user-friendly urban environments.

5.3.3 Integration with Public Transportation

Autonomous robots can be integrated into public transportation systems to enhance accessibility and convenience for passengers. For instance, robots can assist individuals in navigating to and from public transit hubs such as bus stops or train stations, making public transportation a more practical option for those with limited mobility. This integration not only aids individuals requiring additional support but also streamlines passenger flow, improving the overall efficiency of public transportation systems.

One potential application is the use of autonomous shuttles to provide first-mile and last-mile connectivity, transporting passengers between their homes or workplaces and the nearest public transit stations. These shuttles can operate autonomously along predefined routes, offering safe and efficient travel while reducing reliance on private vehicles. Additionally, autonomous robots can assist passengers in navigating large transit hubs, providing directions, and even carrying luggage.

In countries like Japan, autonomous robots have been deployed in train stations to help passengers with wayfinding, provide real-time information on train schedules, and offer language translation services for tourists. By delivering these additional services, autonomous robots contribute to a smoother and more comfortable public transit experience, encouraging increased usage of public transportation and reducing urban traffic congestion.

Overall, the development of autonomous robots for personal and public mobility offers a vision of urban transportation that is more inclusive, efficient, and responsive to the needs of all citizens. By addressing issues related to accessibility and mobility, these robots help create more livable and accommodating urban environments for everyone.

5.4 Commercial and Industrial Sectors

Autonomous robots are playing an increasingly important role in commercial and industrial sectors by significantly improving efficiency, reducing costs, and enhancing scalability. In various fields such as logistics, agriculture, and construction, autonomous robots have become essential tools for meeting growing industry demands while ensuring stability and reliability.

5.4.1 Logistics and Delivery Robots

Autonomous robots are widely adopted in the logistics and delivery sectors to enhance the efficiency of goods movement. Logistics robots play a critical role in warehouses by autonomously navigating aisles and automating tasks such as picking and transporting items between storage and shipping zones. These technologies include solutions like Robotic Arms, Automated Guided Vehicles (AGVs), Automated Storage and Retrieval Systems (AS/RS), and Autonomous Mobile Robots (AMRs), which significantly improve the efficiency of logistics and warehouse management (Dhaliwal, 2020). By automating repetitive tasks, these robots reduce human error, increase throughput, and contribute to cost savings in labor.

Fig. 5.10 Autonomous delivery robot (*Source* https://www.robotis.com/sub/business_012.php) by Robotis, Woowa Brothers (https://robot.baemin.com/)

Additionally, last-mile delivery is another major area where autonomous robots are making a significant impact. Last-mile delivery robots are typically small and compact, designed to navigate sidewalks and city streets to deliver packages directly to consumers' doorsteps. Leading companies such as Starship Technologies and Amazon have developed delivery robots equipped with cameras, sensors, and advanced algorithms to ensure safe navigation and efficient delivery. By replacing traditional delivery vehicles, these robots help reduce traffic congestion and carbon emissions, contributing to the development of more sustainable urban logistics systems.

Companies like ROBOTIS and the Woowa Brothers have developed delivery robots equipped with sensors and algorithms to ensure safe navigation and efficient delivery. By replacing traditional delivery vehicles, these robots reduce traffic congestion and carbon emissions, contributing to the development of a more sustainable urban mule system (Fig. 5.10).

5.4.2 *Agricultural and Construction Robots*

Autonomous robots are transforming agriculture by automating labor-intensive tasks, increasing productivity, reducing costs, and enabling farmers to work more efficiently. Autonomous tractors (Fig. 5.11) and autonomous harvesters (Fig. 5.12) optimize crop yields by using GPS, sensors and AI to precisely navigate farmland. These technologies significantly enhance overall agricultural efficiency by reducing the need for manual labor.

Additionally, autonomous drones are employed for various agricultural purposes, including crop health monitoring, pesticide spraying, and soil condition assessment. These drones provide farmers with real-time data, enabling more informed decision-making. For example, the use of multispectral imaging allows for early detection of water and nutrient deficiencies in crops, contributing to substantial improvements in productivity (Avgoustaki, 2022).

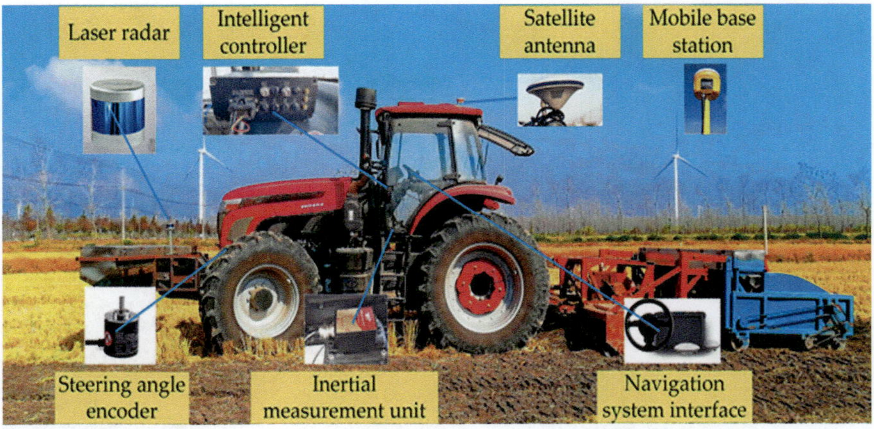

Fig. 5.11 Autonomous tractor (*Source* Qu et al., 2024)

Fig. 5.12 Automated fruit harvesting robot (*Source* Onishi et al., 2019)

Autonomous robots are driving innovative changes not only in agriculture but also in the construction industry. On construction sites, autonomous robots are utilized for various tasks such as bricklaying, material transportation, and site inspection. The semi-autonomous bricklaying robot SAM100 can lay bricks at a much faster rate than human workers, significantly reducing construction time and improving project efficiency (Fig. 5.13).

Additionally, autonomous excavators and bulldozers assist with earthmoving tasks, enhancing both efficiency and safety on construction sites. By automating labor-intensive processes, these construction robots address labor shortages while improving precision and overall project quality.

Fig. 5.13 Bricklaying robot SAM100 (*Source* Dakhli & Lafhaj, 2017)

5.4.3 Manufacturing and Maintenance

In the manufacturing and maintenance sectors, autonomous robots are driving innovation by improving efficiency, precision, and safety across industries. In manufacturing, industrial robotic arms are widely used on assembly lines to perform repetitive and precise tasks such as welding, painting, and assembly (Fig. 5.14). These robots operate continuously without fatigue, ensuring consistent quality and high productivity.

Additionally, collaborative robots (co-bots) work alongside human workers, handling tasks that require precision and flexibility. This collaboration enhances

Fig. 5.14 Industrial robots (*Source* Dzedzickis et al., 2021)

Fig. 5.15 Pedestrian environment assessment using autonomous robots

both productivity and workplace safety, making co-bots an essential component of modern manufacturing processes.

As the importance of pedestrian environment assessment continues to grow, the focus is shifting from traditional subjective methods to objective and efficient approaches using autonomous robots. Equipped with sensors and artificial intelligence, autonomous robots can collect and analyze data, minimizing the need for human resources while delivering highly reliable evaluations. This technology enables national and local governments to efficiently acquire data for improving pedestrian environments (Fig. 5.15).

Autonomous inspection robots are also utilized for monitoring large-scale infrastructure such as pipelines, power lines, and wind turbines. Geo-MMS UAV inspecting railway tracks (Fig. 5.16) while displaying the captured LiDAR point cloud data (Fig. 5.17) used for railway track mapping (Jing et al., 2022). These technologies contribute to optimizing maintenance planning by enabling efficient inspection of large-scale facilities and early detection of defects.

Autonomous robots are being adopted across various industrial sectors, including logistics, agriculture, construction, manufacturing, and maintenance. They set new standards for industry by enhancing efficiency, reducing costs, and improving safety. These examples clearly demonstrate how autonomous robots create value and solve complex problems in diverse industries.

Fig. 5.16 Geo-MMS UAV with Lidar units (*Source* Jing et al., 2022)

Fig. 5.17 Lidar point cloud segments captured for railway mapping (*Source* Jing et al., 2022)

References

Andreasson, H., Magnusson, M., & Lilienthal, A. (2007, October). Has somethong changed here? Autonomous difference detection for security patrol robots. In *2007 IEEE/RSJ International Conference on Intelligent Robots and Systems* (pp. 3429–3435). IEEE.

Avgoustaki, D. D., Avgoustakis, I., Miralles, C. C., Sohn, J., & Xydis, G. (2022). Autonomous mobile robot with attached multispectral camera to monitor the development of crops and detect nutrient and water deficiencies in vertical farms. *Agronomy, 12*(11), 2691.

Basilico, N. (2022). Recent trends in robotic patrolling. *Current Robotics Reports, 3*(2), 65–76.

Bogue, R. (2019). Disaster relief, and search and rescue robots: The way forward. *Industrial Robot: THe International Journal of Robotics Research and Application, 46*(2), 181–187.

Dakhli, Z., & Lafhaj, Z. (2017). Robotic mechanical design for brick-laying automation. *Cogent Engineering, 4*(1), 1361600.

Dhaliwal, A. (2020). The rise of automation and robotics in warehouse management. In *Transforming management using artificial intelligence techniques* (pp. 63–72). CRC Press.

Dzedzickis, A., Subačiūtė-Žemaitienė, J., Šutinys, E., Samukaitė-Bubnienė, U., & Bučinskas, V. (2021). Advanced applications of industrial robotics: New trends and possibilities. *Applied Sciences, 12*(1), 135.

Hirose, N., Tajima, R., & Sukigara, K. (2015, September). Personal robot assisting transportation to support active human life. In *2015 IEEE/RSJ International Conference on Intelligent Robots and Systems (IROS)* (pp. 5395–5402). IEEE.

Jing, G., Qin, X., Wang, H., & Deng, C. (2022). Developments, challenges, and perspectives of railway inspection robots. *Automation in Construction, 138*, 104242.

Kumar, V. V., Shrimali, M., Shaik, N., Ranjitha, K. N., Garg, N., & Maranan, R. (2024, October). Navigating the dark: Advances in robotic night patrol with D-block mask electric EEL dense nested R-CNN for enhanced safety. In *2024 4th International Conference on Sustainable Expert Systems (ICSES)* (pp. 1034–1041). IEEE.

Mahalakshmi, R., Kavitha, M., Gopi, B., & Kumar, S. M. (2023, January). Women safety night patrolling IoT robot. In *2023 5th International Conference on Smart Systems and Inventive Technology (ICSSIT)* (pp. 544–549). IEEE.

Memmesheimer, R., Overbeck, M., Kral, B., Steffen, L., Behnke, S., Gersch, M., & Roennau, A. (2024). *Cleaning robots in public spaces: A survey and proposal for benchmarking based on stakeholders interviews.* arXiv preprint arXiv:2407.16393.

Onishi, Y., Yoshida, T., Kurita, H., Fukao, T., Arihara, H., & Iwai, A. (2019). An automated fruit harvesting robot by using deep learning. *Robomech Journal, 6*(1), 1–8.

Qu, J., Zhang, Z., Qin, Z., Guo, K., & Li, D. (2024). Applications of autonomous navigation technologies for unmanned agricultural tractors: A review. *Machines, 12*(4), 218.

Ruan, K., Wu, Z., & Xu, Q. (2021). Smart cleaner: A new autonomous indoor disinfection robot for combating the covid-19 pandemic. *Robotics, 10*(3), 87.

SeungSub, O., Jehun, H., Hyunjung, J., Soyeon, L., & Jinho, S. (2017, June). A study on the disaster response scenarios using robot technology. In *2017 14th International Conference on Ubiquitous Robots and Ambient Intelligence (URAI)* (pp. 520–523). IEEE.

Simondetti, A., Bachand, N., Delahunty, A., Griffith, J., & Sustarevas, J. (2021). *Autonomous robotics in the AEC practice.*

Vorina, A., Ojsteršek, T., & Pušnik, D. (2022). Autonomous delivery robots and their contribution during the pandemic. *Horizons-International Scientific Journal, 31*(2), 79–86.

Wang, H., Grindle, G. G., Candiotti, J., Chung, C., Shino, M., Houston, E., & Cooper, R. A. (2012, August). The personal mobility and manipulation appliance (PerMMA): A robotic wheelchair with advanced mobility and manipulation. In *2012 Annual International Conference of the IEEE Engineering in Medicine and Biology Society* (pp. 3324–3327). IEEE.

Wang, X. V., & Wang, L. (2021). A literature survey of the robotic technologies during the COVID-19 pandemic. *Journal of Manufacturing Systems, 60*, 823–836.

Ye, X., & Robert, L. P. (2023, August). Human security robot interaction and anthropomorphism: An examination of pepper, RAMSEE, and Knightscope robots. In *2023 32nd IEEE International Conference on Robot and Human Interactive Communication (RO-MAN)* (pp. 982–987). IEEE.

Chapter 6
Integration of Autonomous Robots in Environmental and Urban Planning

Autonomous robots have emerged as a pivotal technological innovation capable of transforming urban life (Woo et al., 2020). These robots are more than just mechanical devices—they are key elements that can significantly impact urban mobility, convenience, and sustainability. By redefining interactions between pedestrians, vehicles, and public transportation, autonomous robots have the potential to enhance the quality of life for city residents while contributing to the efficient restructuring of urban functions. Their integration into public transportation, pedestrian-friendly initiatives, and urban planning efforts highlights their crucial role in improving accessibility and mobility.

The introduction of autonomous robots has the potential to fundamentally reshape not only technological advancements but also the structure and functionality of urban environments. Modern cities are grappling with persistent challenges such as traffic congestion and environmental degradation, and autonomous robots can play a key role in addressing these issues. For instance, by automating last-mile logistics deliveries in urban centers, these robots can help alleviate road congestion and reduce carbon emissions (Kotlars et al., 2024). Additionally, they can be deployed for various public services, such as pedestrian assistance, street cleaning and maintenance, and patrolling, significantly enhancing the quality of urban life. These advancements are particularly beneficial for vulnerable populations, including the elderly, in an aging society.

Autonomous robots can also contribute to the transformation of urban pedestrian environments, facilitating the shift toward pedestrian-centric cities. As many cities seek to move away from vehicle-dominated transportation models, autonomous robots can enhance pedestrian convenience and improve safety on walkways. For example, they can help maintain pedestrian-friendly environments by automatically removing obstacles from sidewalks or assisting pedestrians in safely crossing streets (Choi et al., 2024). By improving the quality of pedestrian spaces, these robots can encourage walking as a more attractive mode of transport, ultimately influencing broader urban mobility patterns.

M. Choi and S. Lee, *Intelligent Moving Cities: Technological Leap and Social Integration of Autonomous Mobile Robots*,
https://doi.org/10.1007/978-981-96-8864-7_6

Furthermore, autonomous robots can play a significant role in enhancing urban mobility by integrating with public transportation systems. Addressing the last-mile challenge—connecting public transit stops with final destinations, these robots can improve accessibility and convenience, encouraging more people to choose public transportation over private vehicles. For instance, they can assist passengers at transit stops, transporting them safely to their final destinations. This service is particularly beneficial for elderly individuals or people with disabilities who may face challenges in walking, thereby increasing public transportation usage across urban areas. Additionally, by synchronizing in real-time with public transit systems, autonomous robots can help alleviate traffic congestion and maximize transportation efficiency.

This chapter explores the multifaceted impact of autonomous robots on urban environments and the transformations they bring. First, it examines both the benefits and challenges of introducing autonomous robots in cities, highlighting how they can improve urban life and support the transition toward pedestrian-friendly urban spaces. Next, the chapter discusses how autonomous robots interact with public transportation, enhancing convenience and accessibility while addressing the last-mile problem.

Finally, policy, industrial, and urban planning strategies are proposed to ensure the successful integration of autonomous robots into urban and transportation systems. The widespread adoption of these technologies requires collaboration among various stakeholders to promote the social, economic, and environmental sustainability of cities (Gavino et al., 2023). To this end, the chapter presents an in-depth discussion on the role of autonomous robots in urban planning, the policy measures needed for their integration, and the potential for industrial development. By outlining strategies for incorporating autonomous robots into the fabric of future cities, this chapter provides a comprehensive perspective on how they can harmoniously interact with pedestrians, vehicles, and public transportation.

Ultimately, autonomous robots represent not just an advancement in technology but a crucial key to realizing human-centered cities. By contributing to safer, more convenient, and more sustainable urban living, these robots will play an essential role in shaping the future of cities. With a focus on sustainable urban development and improving the quality of life for citizens, their integration into urban environments will pave the way for smarter, more efficient, and more inclusive cities.

6.1 Introduction of Autonomous Robots into Urban Environments and Their Impact

Modern cities face numerous challenges due to population growth and urbanization (Echendu et al., 2021). To address various issues such as traffic congestion, environmental pollution, and a lack of convenience in daily life, cities are adopting new technologies, among which autonomous robots are a key example. Autonomous robots are not simply technological innovations; they transform multiple aspects of urban life and facilitate the transition towards pedestrian-centric cities. This section aims to explore the diverse impacts arising from the introduction of autonomous robots into urban areas and to discuss how they may alter the lives of urban residents.

6.1.1 *The Significance of Introducing Autonomous Robots into Urban Areas*

The introduction of autonomous robots into urban environments holds significance beyond mere technological advancement. It marks a critical turning point toward human-centered urban development, introducing various potential transformations as these robots become part of city spaces (Galati et al., 2024). Autonomous robots serve diverse urban functions, including small-scale logistics delivery, security surveillance, and public service provision, significantly enhancing urban efficiency and safety. As illustrated in Fig. 6.1, autonomous robots provide services in various urban settings, including security, delivery, medical assistance, and guidance roles.

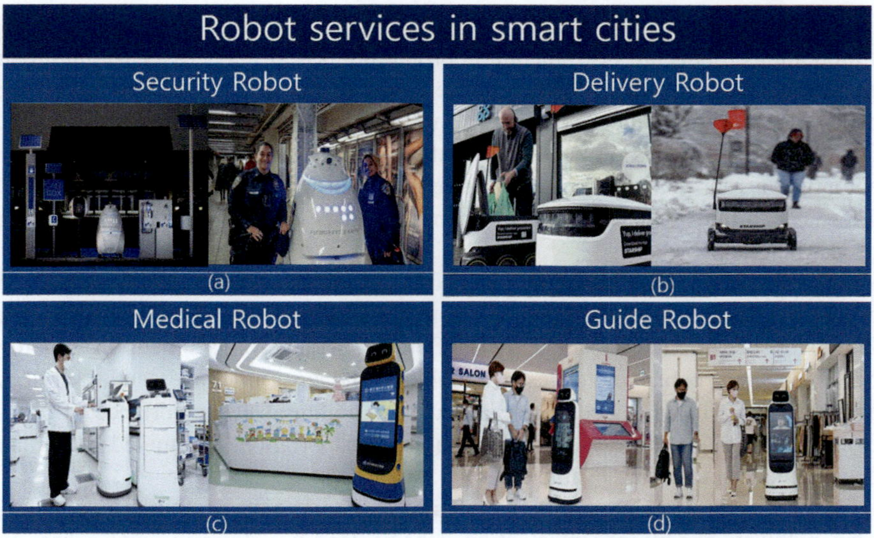

Fig. 6.1 Robot services in smart cities. Security robots, as shown in **a**, are deployed in public spaces to patrol and monitor for potential security threats. This image is sourced from Knightscope (https://www.knightscope.com/use-cases/law-enforcement). Delivery robots, depicted in **b**, are used for logistics and food delivery within smart cities, enhancing efficiency in urban environments. This image is sourced from Starship Technologies (https://www.starship.xyz/). Medical robots, represented in **c**, such as LG CLOi, perform tasks like visitor guidance, medical supply delivery, and nighttime patrols within hospitals, offering personalized services. This image is sourced from LG Electronics (https://live.lge.co.kr/lg-cloi-2/). Guide robots, shown in **d**, assist users in navigating complex spaces such as shopping malls and airports, improving accessibility and convenience. This image is sourced from LG Electronics (https://live.lge.co.kr/lg-cloi-guidebot/)

In addition to these applications, autonomous robots are designed to execute novel functions while seamlessly integrating with pre-existing urban infrastructure. For instance, these robots can navigate the boundary between roads and pedestrian pathways, facilitating goods transport and assisting in scenarios where pedestrian mobility is hindered. This capability proves particularly advantageous in metropolitan areas with heavy traffic congestion. By optimizing last-mile logistics delivery, autonomous robots contribute to reducing the number of vehicles on urban roads, thereby enhancing the pedestrian environment.

Moreover, autonomous robots play a crucial role in strengthening urban security and safety (Huang et al., 2019). These robots can perform autonomous patrol duties in public spaces, monitor potential risk situations, and respond promptly when necessary. As a result, urban residents can experience a safer living environment, while law enforcement and security personnel can allocate resources more efficiently. The ability of autonomous robots to augment public safety systems fosters an urban environment where pedestrians can move freely with a heightened sense of security.

6.1.2 Enhancement of Urban Quality of Life

Autonomous robots are instrumental in promoting pedestrian activity within urban environments. Improvements in the pedestrian environment contribute to making pedestrian movement more convenient and safer for citizens (Valdez et al., 2023). For instance, obstacles on pedestrian pathways can be automatically removed by autonomous robots, and pedestrians can receive assistance in safely crossing streets. These capabilities enhance the pedestrian experience and contribute to promoting pedestrian mobility.

In recent years, urban researchers and policymakers have increasingly emphasized pedestrian-centric urban planning to create more sustainable and accessible cities. Initiatives such as Paris's 15-min city and Barcelona's superblocks focus on reducing car dependency, enhancing pedestrian mobility, and ensuring that essential services are within walking distance. These approaches highlight the importance of high-quality pedestrian pathways, seamless public transportation, and mixed-use urban development in fostering livable cities.

As part of this growing focus on walkability, studies have examined how pedestrian pathway conditions, intersection density, land-use diversity, and access to public transportation influence pedestrian behavior and mobility patterns. These research efforts have led to the development of walkability assessment frameworks, which provide structured methodologies for evaluating the quality of pedestrian environments.

Jeong et al. (2023) introduced the Walkability Evaluation System (WES) as a framework to assess pedestrian environments based on key factors such as pedestrian pathway length, intersection density, and land-use diversity. This system highlights the critical elements that contribute to pedestrian-friendly urban spaces. Autonomous robots can enhance these factors by actively monitoring pedestrian pathways, identifying damages, and facilitating quick maintenance, ensuring that pathways remain accessible and safe for all users.

To actively improve walkability, autonomous robots can play a crucial role by conducting real-time monitoring of pedestrian pathways. By detecting damages, identifying obstacles, and reporting necessary repairs, these robots help maintain an optimal walking environment. This not only enhances pedestrian safety but also encourages more people to choose walking as a primary mode of transport.

Additionally, autonomous robots can significantly enhance accessibility for mobility-impaired individuals. Real-time assessments of pathway inclines, surface conditions, and temporary obstructions enable robots to provide personalized navigation assistance for wheelchair users and other individuals with mobility challenges. Through adaptive route recommendations, robots ensure barrier-free mobility options, making urban environments more inclusive.

Beyond individual assistance, autonomous robots can integrate with smart city infrastructure to create a dynamic walkability management system. By continuously evaluating pedestrian environments, robots can provide urban planners with real-time data on maintenance needs, pedestrian density, and accessibility concerns. This allows city authorities to optimize pedestrian routes and improve urban mobility planning efficiently.

Autonomous robots can also be utilized in various ways within urban public spaces. For instance, in public parks, autonomous robots can perform maintenance tasks to ensure that the parks remain clean and well-maintained or assist children and the elderly in navigating these spaces. This contributes to increasing the utilization of urban parks and providing citizens with a better recreational environment. Additionally, autonomous robots can play a role in public events or festivals by providing guidance, security monitoring, or logistical support, thereby enhancing the efficiency of events and improving convenience for participants.

Finally, the introduction of autonomous robots can play a crucial role in improving urban traffic flow and ensuring pedestrian safety in vehicle-dense areas. By monitoring both pedestrian and vehicle movements in real time, autonomous robots help reduce congestion and minimize the risk of accidents. Furthermore, robots can assist in traffic-calming strategies, which are critical for ensuring that streets are designed to prioritize pedestrian movement over high-speed vehicle traffic. Ultimately, this contributes to creating an environment where urban residents can move more safely and efficiently.

The introduction of autonomous robots serves as a catalyst for transitioning urban environments towards a pedestrian-centric model. Autonomous robots enhance pedestrian convenience, improve the quality of urban life, and promote sustainable urban development by encouraging pedestrian activity and integrating with public transportation. These transformations not only make urban life more convenient and safer but also contribute to enhancing the environmental and social sustainability of cities.

6.2 Interaction Between Public Transportation and Autonomous Robots

Public transportation serves as a crucial foundation for supporting the daily mobility of citizens in modern urban environments (Joshi et al., 2021). However, public transportation systems in many cities still face challenges such as limited accessibility, first-mile and last-mile issues, and congestion. To address these challenges, the introduction of autonomous robots offers an innovative solution through interactions with public transportation. This chapter aims to discuss how autonomous robots can be integrated with public transportation to enhance urban mobility efficiency and improve the travel experience for citizens.

6.2.1 Role of Autonomous Robots in Promoting Public Transportation

Autonomous robots can enhance the efficiency and convenience of public transportation systems through their interactions with these networks. The journey from public transportation stops to final destinations is often perceived as inconvenient, which is one of the factors discouraging people from using public transportation (Gurumurthy et al., 2020).

To address this issue, autonomous robots can improve the accessibility of public transportation by providing last-mile services. For instance, autonomous robots can pick up passengers from public transportation stops and transport them safely to their final destinations. This service is particularly beneficial for elderly individuals or people with disabilities who face challenges with walking. Additionally, autonomous robots can be integrated with public transportation systems to provide optimal routes based on real-time traffic information, thereby minimizing potential inconveniences that passengers may encounter when using public. Figure 6.2 illustrates the integration of ADRs with public transportation as part of a smart city logistics solution (De Maio et al., 2024). Such an approach encourages urban residents to use public transportation more conveniently and contributes to reducing dependence on private vehicles.

Furthermore, autonomous robots can fill gaps in public transportation services. For example, during late-night hours or periods of low public transportation demand, autonomous robots can serve as shuttle services, transporting passengers to specific areas. This not only enhances the operational efficiency of public transportation but also provides citizens with a 24-h mobility solution, thereby increasing public trust in the overall transit system.

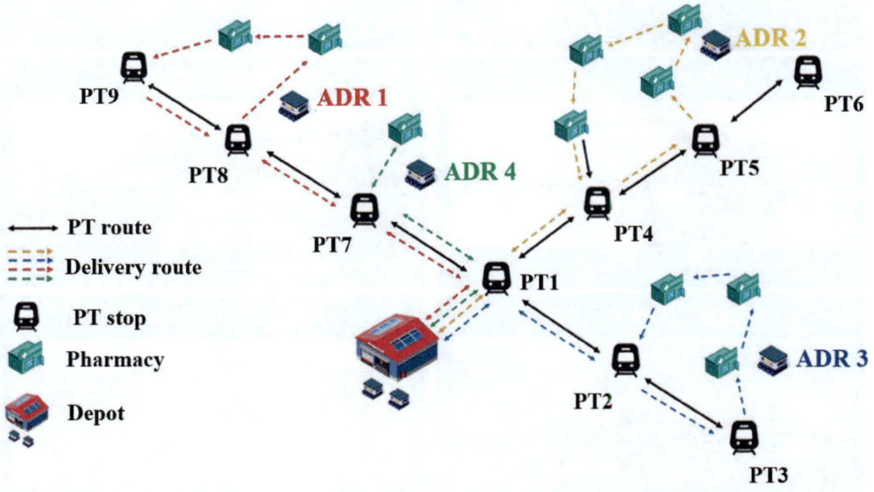

Fig. 6.2 Autonomous Delivery Robot (ADR) routing with public transportation. Schematic of ADRs integrated with public transportation for smart city logistics, enhancing delivery range, efficiency, and sustainability (De Maio et al., 2024)

6.2.2 Mobility Hubs and Autonomous Systems

Mobility hubs are centralized urban spaces where various modes of transportation converge, providing seamless transitions for citizens between different transport options (Ku et al., 2022). As depicted in Fig. 6.3, these hubs connect first-mile and last-mile travel through integrated services such as bike sharing, car sharing, charging stations, and logistics facilities, enhancing overall mobility efficiency. By strategically integrating public and shared mobility, these hubs reduce dependency on private vehicles and promote sustainable urban transport.

To accommodate diverse urban mobility needs, mobility hubs are designed in various forms based on their scale and function. As shown in Fig. 6.4, mobility hubs can be categorized into four types: community hubs, metabolism hubs, public

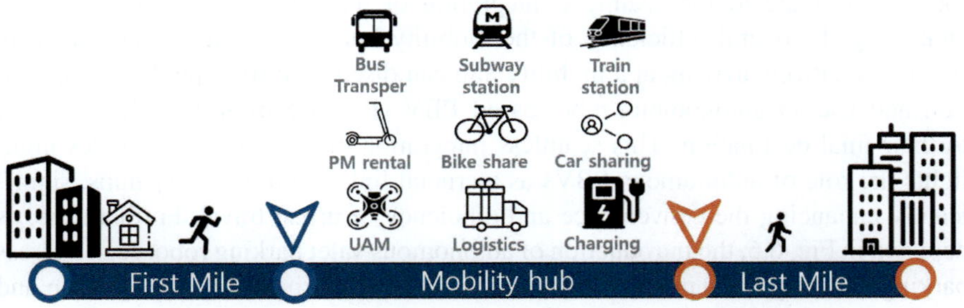

Fig. 6.3 Mobility hub flow

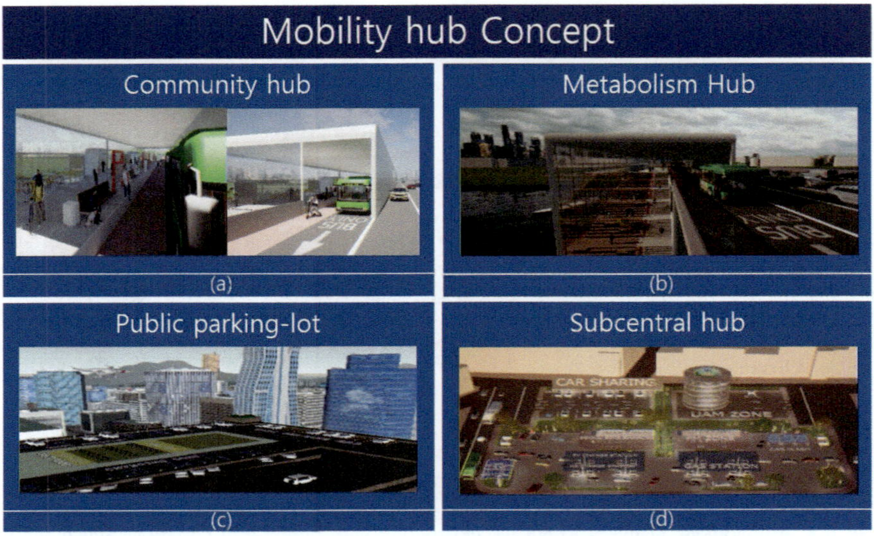

Fig. 6.4 Concept of mobility hub

parking-lot hubs, and subcentral hubs. Community hubs enhance last-mile connectivity by integrating bus stops and shared mobility services. Metabolism hubs facilitate multimodal transport by linking various transportation modes in key urban areas. Public parking-lot hubs connect parking facilities with public transit to encourage shifts away from private vehicle use. Subcentral hubs serve as regional transfer points, integrating multiple transport services such as gas and electric charging stations, personal mobility (PM) parking, car-sharing services, and urban air mobility (UAM) infrastructure. By providing seamless multimodal connections, subcentral hubs enhance accessibility between suburban and central areas, further improving the efficiency of urban transportation networks.

Autonomous robots can play a significant role within these mobility hubs. Mobility hubs bring together diverse transportation modes such as public transportation, bike-sharing systems, and personal mobility, creating an environment where users can conveniently switch between different means of transport. Autonomous robots contribute to the seamless integration of these transport modes, thereby enhancing the overall efficiency of the mobility hubs. For instance, as shown in Fig. 6.5, a citizen arriving at a mobility hub can disembark from public transportation and use an autonomous robot-based PBV (Purpose-Built Vehicle) to travel to their final destination. This seamless integration of transportation modes highlights the role of autonomous PBVs as a crucial link within mobility hubs, significantly enhancing the convenience and efficiency of urban travel. Furthermore, as depicted in Fig. 6.6, the introduction of autonomous valet parking robots can address parking challenges for private vehicle users, reducing urban traffic congestion and encouraging a transition towards increased reliance on public transportation.

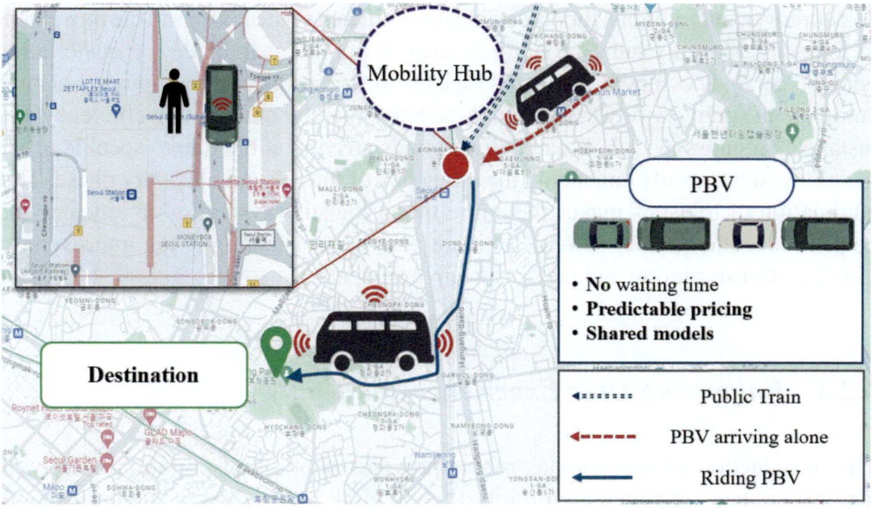

Fig. 6.5 Integration of autonomous robots in mobility hubs

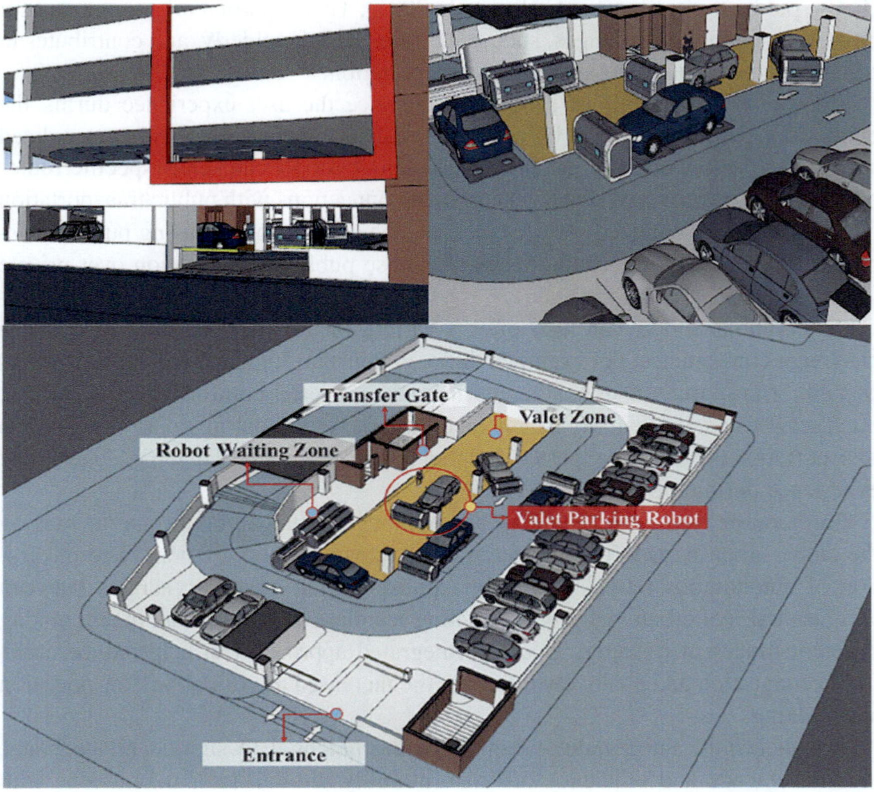

Fig. 6.6 Autonomous valet parking robot system

Moreover, autonomous robots can make significant contributions to the movement of logistics. Mobility hubs are not only critical points for individual transportation but also serve as essential nodes for logistics distribution. Autonomous robots can automate the delivery process within these hubs, thereby maximizing efficiency. For instance, autonomous robots can deliver goods arriving at the hub to specific urban locations, significantly enhancing the efficiency of the last-mile delivery phase. This automation of logistics makes the entire urban logistics system more efficient and helps reduce the number of vehicles in city areas, contributing to the improvement of public transportation and pedestrian environments.

6.2.3 Enhancing User Experience and Utility

The introduction of autonomous robots provides public transportation users with new experiences and increased utility (Sahoo et al., 2023). By facilitating smooth travel between public transportation stops and final destinations, autonomous robots improve the accessibility of public transportation, making it easier for citizens to choose public transportation (Ceder et al., 2021). This is particularly significant for those with mobility challenges, such as children and the elderly, and contributes to expanding the inclusiveness of public transportation.

Additionally, autonomous robots can enhance the user experience during the public transportation journey by providing real-time information and personalized services to passengers. For instance, autonomous robots can select specific routes based on passenger needs or suggest optimal connections with public transportation options. Such personalized services increase the convenience of using public transportation, positively influencing users to choose public transportation over private vehicles.

Autonomous robots can also offer a variety of auxiliary services to enhance passenger convenience. For example, they can transport luggage or greet passengers at public transportation stops, thereby increasing overall comfort and ease of use. Such services are particularly beneficial during peak commuting hours when public transportation is heavily congested, helping to reduce the stress that passengers may experience during their journey.

Furthermore, the introduction of autonomous robots can contribute to making the entire urban transportation system more integrated and efficient. The integration of autonomous robots with public transportation creates synergies between various modes of urban mobility, facilitating seamless transitions for citizens among different modes of transport. Such an integrated approach can help reduce urban traffic congestion and positively influence the increased use of public transportation and walking.

In conclusion, autonomous robots have the potential to significantly enhance the convenience and efficiency of urban mobility through their interactions with public transportation. The use of autonomous robots addresses the last-mile problem, provides seamless connections at mobility hubs, and offers personalized services to passengers. As a result, autonomous robots can promote the use of public transportation and improve the mobility experience for urban residents. Ultimately, this approach reduces dependence on private vehicles and facilitates the transition towards a sustainable urban environment centered on walking and public transportation.

6.3 Strategies for Integrating Urban Planning and Autonomous Robots

The introduction of autonomous robots into urban environments requires a multifaceted approach involving policy, social, and technical perspectives to ensure effective integration (Yigitcanlar et al., 2019). To operate safely and efficiently within cities, autonomous robots necessitate collaboration and consensus among various stakeholders, supported by comprehensive urban planning and policy frameworks. This chapter aims to discuss the establishment of policy agreements, planning strategies related to the adoption of autonomous robots, potential industrial implications, and approaches for integration with urban planning.

6.3.1 Policy Consensus and Planning

Successful adoption of autonomous robots requires collaboration and policy consensus among various stakeholders (Ekatpure et al., 2024). For autonomous robots to operate smoothly in urban areas, changes to urban infrastructure, including roads and pedestrian pathways, are essential, and these changes are closely linked to urban planning. As illustrated in Fig. 6.7, changes in urban spatial structure with the introduction of autonomous robot services demonstrate how urban spaces evolve to accommodate seamless integration of various robotic mobility solutions, such as autonomous shuttles, delivery robots, and pedestrian-friendly zones. These services operate within multimodal pathways, optimizing space utilization and ensuring efficient transportation for diverse user needs. Government bodies, local municipalities, private companies, and citizen organizations must come together to discuss the roles and integration strategies of autonomous robots, subsequently establishing clear legal regulations and operational standards based on these discussions.

In particular, regulations to address potential safety issues caused by the introduction of autonomous robots are crucial. To mitigate risks associated with interactions between autonomous robots and pedestrians, clear regulations are required, including speed limitations for robots, route planning, and protocols for managing emergency situations (Winfield et al., 2018).

Additionally, data management and privacy protection are critical considerations in the adoption of autonomous robots (Khan et al., 2020). For autonomous robots to operate smoothly in urban environments, real-time data collection and analysis are

Fig. 6.7 Changes in urban spatial structure with the introduction of autonomous robot services (https://www.hyundai.com/worldwide/en/newsroom/detail/hyundai-motor-honored-with-inaugural-autoweek-vanguard-award-0000000324)

essential. In this process, policy measures must be established to address potential privacy violations that could arise from the collection and use of personal information.

6.3.2 Industrial Potential of Autonomous Robots

The introduction of autonomous robots into urban environments can drive industrial transformation by creating new economic opportunities and fostering technological advancements (Cone et al., 2019). These robots have the potential to generate employment across various sectors, including the development, manufacturing, and maintenance of autonomous robotics technologies, thereby stimulating local economic growth.

Moreover, the adoption of autonomous robots is expected to redefine operations in multiple service industries, particularly logistics, security, and cleaning, by driving automation and efficiency. For instance, autonomous robots can enhance logistics operations within urban areas by automating transportation and warehouse management, significantly enhancing delivery speed and cost-efficiency, as shown in Fig. 6.8. By streamlining last-mile delivery, these robots reduce operational costs and delivery time, thereby strengthening business competitiveness in urban logistics.

Beyond logistics, autonomous robots play a crucial role in security and urban patrolling. By conducting real-time monitoring in urban public spaces, these robots can supplement security personnel, reduce human resource strain, and enhance overall urban safety. Their ability to operate continuously and autonomously makes

Fig. 6.8 Autonomous Robots in the Logistics Industry. This figure highlights the application of autonomous robots transforming the logistics industry. Panel **a**, sourced from Locus Robotics (https://locusrobotics.com/), shows mobile robots optimizing warehouse operations by efficiently navigating and transporting items. Panel **b**, sourced from Grey Orange (https://www.greyorange.com/), illustrates an automated storage and retrieval system (ASRS), where robots enhance storage efficiency and accelerate item retrieval, demonstrating cutting-edge goods-to-person automation

them valuable in high-traffic areas, transportation hubs, and commercial districts, where ensuring security is paramount.

In addition to improving established industries, the deployment of autonomous robots presents new business opportunities for startups and emerging technology companies. Innovations in autonomous driving, artificial intelligence, and robotics enable companies to develop novel applications and services, further contributing to urban development and smart city infrastructure (Khan et al., 2024). For instance, startups can leverage autonomous robots for services such as automated food delivery, intelligent surveillance, and AI-driven maintenance systems, offering solutions to urban challenges.

This demonstrates that the integration of autonomous robots is not merely a technological shift but a significant driver of economic growth and industrial evolution. As cities continue to adopt smart infrastructure and automation, the role of autonomous robots will expand, reshaping the urban economy, improving service efficiency, and driving sustainable industrial development.

6.3.3 Integration with Urban Planning

For autonomous robots to be successfully integrated into urban environments, strategic urban planning and close coordination among stakeholders are essential. Urban planning provides the necessary infrastructure for autonomous systems while ensuring that these robots can operate safely and efficiently (Fadhel et al., 2024). This includes dedicated pathways for autonomous robots, well-designed intersections for safe pedestrian-robot interaction, and multimodal transfer centers that accommodate a variety of transportation technologies.

Fig. 6.9 Vision for the Gwangmyeong-Siheung mobility hub city

A notable example of this approach can be found in Korea's Gwangmyeong-Siheung mobility planning district. As part of its broader vision to develop a super-connected mobility hub, the district is actively integrating autonomous robots, shuttles, and other advanced mobility services. This initiative involves building road infrastructure optimized for autonomous systems, designing pedestrian-friendly pathways, and creating multimodal transit hubs to support seamless connectivity. As illustrated in Fig. 6.9, the district's infrastructure includes dedicated lanes for autonomous vehicles, pedestrian-friendly pathways, and smart transit stations, ensuring an efficient and sustainable urban transport ecosystem. Additionally, Urban Air Mobility (UAM) vehicles, such as flying taxis and cargo drones, highlight the district's commitment to future-oriented mobility solutions. This comprehensive approach demonstrates how well-planned infrastructure can improve urban mobility, enhance sustainability, and support the large-scale deployment of autonomous technologies.

In conclusion, the successful integration of autonomous robots into urban environments requires comprehensive urban planning, strategic policymaking, and industrial collaboration. By leveraging well-designed infrastructure, public engagement, and sustainability-focused policies, autonomous robots can enhance urban mobility, reduce congestion, and support public transportation and pedestrian-centric development (Noyman et al., 2017). The case of Gwangmyeong-Siheung illustrates how forward-thinking urban planning can facilitate these advancements, creating a model for future smart cities. To fully realize the potential of autonomous robots, stakeholders across sectors must collaborate to establish the necessary regulatory frameworks and urban conditions for successful deployment.

References

Ceder, A. (2021). Urban mobility and public transport: Future perspectives and review. *International Journal of Urban Sciences, 25*(4), 455–479.

Choi, M. J., Park, S. J., Kim, S., & Lee, S. J. (2024). Human-aware trajectory optimization for enhancing D* Algorithm for autonomous robot navigation (May 2024). *IEEE Access*.

Cone, E., & Lambert, J. (2019). *How robots change the world*.

De Maio, A., Ghiani, G., Laganà, D., & Manni, E. (2024). Sustainable last-mile distribution with autonomous delivery robots and public transportation. *Transportation Research Part c: Emerging Technologies, 163*, 104615.

Echendu, A. J., & Okafor, P. C. C. (2021). Smart city technology: A potential solution to Africa's growing population and rapid urbanization? *Development Studies Research, 8*(1), 82–93.

Ekatpure, R. (2024). Safety protocols and risk mitigation strategies in the implementation of autonomous driving systems. *Advances in Urban Resilience and Sustainable City Design, 16*(02), 37–46.

Fadhel, M. A., Duhaim, A. M., Saihood, A., Sewify, A., Al-Hamadani, M. N., Albahri, A. S., ... Gu, Y. (2024). Comprehensive systematic review of information fusion methods in smart cities and urban environments. *Information Fusion, 107*, 102317.

Galati, G. (2024). *Human-centered robotics: challenges in socially aware navigation and mission planning*.

Gavino, S., & Sobhani, S. E. (2023). *Fostering mission-oriented innovation and AI for sustainable cities and communities: Advancing collaboration and synergistic progress towards the 2030 agenda*.

Gurumurthy, K. M., Kockelman, K. M., & Zuniga-Garcia, N. (2020). First-mile-last-mile collector-distributor system using shared autonomous mobility. *Transportation Research Record, 2674*(10), 638–647.

Huang, L., Zhou, M., Hao, K., & Hou, E. (2019). A survey of multi-robot regular and adversarial patrolling. *IEEE/CAA Journal of Automatica Sinica, 6*(4), 894–903.

Jeong, I., Choi, M., Kwak, J., Ku, D., & Lee, S. (2023). A comprehensive walkability evaluation system for promoting environmental benefits. *Scientific Reports, 13*(1), 16183.

Joshi, S., Bailey, A., & Datta, A. (2021). On the move? Exploring constraints to accessing urban mobility infrastructures. *Transport Policy, 102*, 61–74.

Khan, I. U., Taherdoost, H., Madanchian, M., Ouaissa, M., El Hajjami, S., & Rahman, H. (Eds.). (2024). *Future tech startups and innovation in the age of AI*. CRC Press.

Khan, M. S. N. (2020). *Privacy in the age of autonomous systems* (Doctoral dissertation, KTH Royal Institute of Technology).

Kotlars, A., & Skribans, V. (2024). Literature review: Efficiency, environment and robotization in first and last mile logistics. *Transportation Research Interdisciplinary Perspectives, 27*, 101215.

Ku, D., Choi, M., Lee, D., & Lee, S. (2022). The effect of a smart mobility hub based on concepts of metabolism and retrofitting. *Journal of Cleaner Production, 379*, 134709.

Noyman, A., Stibe, A., & Larson, K. (2017). *Roadmap for Autonomous Cities: Sustainable Transformation of Urban Spaces*.

Sahoo, S. K., & Choudhury, B. B. (2023). A review on human-robot interaction and user experience in smart robotic wheelchairs. *Journal of Technology Innovations and Energy, 2*(2), 39–55.

Valdez, M., & Cook, M. (2023). Humans, robots and artificial intelligences reconfiguring urban life in a crisis. *Frontiers in Sustainable Cities, 5*, 1081821.

Winfield, A. F., & Jirotka, M. (2018). Ethical governance is essential to building trust in robotics and artificial intelligence systems. *Philosophical Transactions of the Royal Society a: Mathematical, Physical and Engineering Sciences, 376*(2133), 20180085.

Woo, J., Whittington, J., & Arkin, R. (2020). Urban robotics: Achieving autonomy in design and regulation of robots and cities. *Connecticut Law Review, 52*, 319.

Yigitcanlar, T., Wilson, M., & Kamruzzaman, M. (2019). Disruptive impacts of automated driving systems on the built environment and land use: An urban planner's perspective. *Journal of Open Innovation: Technology, Market, and Complexity, 5*(2), 24.

Chapter 7
Challenges and Prospects for Autonomous Robots

The integration of autonomous robots into urban environments represents a significant societal shift, one that extends beyond technological advancements to encompass complex social, ethical, and regulatory considerations. Unlike conventional technologies, autonomous robots interact directly with human surroundings, necessitating careful attention to their societal impact, ethical implications, and governance frameworks. As cities increasingly adopt autonomous solutions, it is crucial to ensure their responsible integration in ways that align with social values, public expectations, and broader urban development goals.

Autonomous robots have the potential to transform urban life by improving efficiency, enhancing safety, and introducing new mobility solutions. However, their successful adoption requires addressing challenges that go beyond technical innovation (Abbass, 2019). Key concerns include ethical decision-making in critical situations, data privacy, public trust, and the establishment of a robust regulatory framework. The introduction of autonomous robots is not solely a matter of technological readiness; it also requires societal adaptation and acceptance. A balanced approach—one that considers both technological progress and human factors—is essential to ensure these systems are beneficial and widely embraced.

This chapter explores the multifaceted challenges associated with integrating autonomous robots into human-centric environments. It delves into technical and ethical concerns (Sect. 7.1), examines the legal and regulatory frameworks required for responsible deployment (Sect. 7.2), and considers long-term perspectives and social acceptance (Sect. 7.3). By addressing these key aspects, this chapter aims to provide a comprehensive understanding of the complexities involved in introducing autonomous robots into urban settings while offering insights into how these challenges can be navigated to create more inclusive, safe, and well-integrated urban environments.

Section 7.1 discusses the technical and ethical challenges autonomous robots face, including ethical decision-making in unpredictable scenarios, data privacy concerns, and the crucial role of public trust.

M. Choi and S. Lee, *Intelligent Moving Cities: Technological Leap and Social Integration of Autonomous Mobile Robots*,
https://doi.org/10.1007/978-981-96-8864-7_7

Section 7.2 examines the regulatory and legal considerations essential for the safe deployment of autonomous robots, highlighting the importance of clear regulatory standards, liability issues, and the intersection of ethical principles and legal frameworks.

Section 7.3 explores long-term perspectives and public acceptance, emphasizing the role of education, community engagement, and societal integration. It underscores the need for transparency, active public participation, and collaboration to build social trust and ensure the sustained success of these technologies.

By systematically analyzing these challenges, this chapter aims to outline a path for the responsible and ethical deployment of autonomous robots—one that maximizes their potential benefits while addressing the concerns of all stakeholders involved.

7.1 The Technical and Ethical Challenges

7.1.1 The Ethical Dilemma of Autonomous Decision-Making

Autonomous robots must make decisions in real time in complex and unpredictable environments. These decisions can have a significant impact on human safety and well-being, which can lead to difficult ethical dilemmas. For example, in an accident scenario, an autonomous robot must choose between the safety of passengers, pedestrians, or passersby. Figure 7.1 shows the dilemma situation. These questions are not only technical but also ethical, requiring developers to design decision-making algorithms that consider safety and fairness. This subsection examines various ethical frameworks, including utilitarianism and deontology, and how these frameworks can be applied to the decision-making process of autonomous robots.

Ethical issues are further complicated by the fact that ethical norms vary from society to society and from culture to culture. What is considered ethically acceptable in one culture can be problematic in another (Pekerti & Arli, 2017). For example, a decision-making algorithm designed to minimise overall damage may be unfair or discriminatory for a particular group. Therefore, developers and policymakers must consider this cultural context and strive to create systems that respect local values and norms while adhering to universal ethical principles. The ethical implications of autonomous decision-making are not limited to accident scenarios. Autonomous robots also make decisions in everyday situations, such as determining the most efficient route or how to interact with pedestrians. These everyday decisions can have broader social implications if the algorithms used are biased or inadvertently prioritize the needs of one group over another. This subsection also discusses the importance of algorithms that are designed to promote fairness and equity, and are free from bias in their decision-making algorithms.

Fig. 7.1 Ethical decision of autonomous robot

7.1.2 Protection of Personal Information and Data

In particular, the collection of data by autonomous robots in public places raises serious privacy concerns. Robots rely on sensors, cameras, and other data-gathering tools to navigate their environment and perform their functions. This results in the continuous collection of vast amounts of data, including potentially sensitive personal information such as images of individuals, location data, and even biometric information. Ensuring that this data is handled in compliance with privacy laws and ethically is a significant challenge (Farhud & Zokaei, 2021). Figure 7.2 shows the collection of object images that the robot detects, and it needs to obscure certain elements to maintain privacy. Human faces and car license plates need to be blurred, and the collected data must be encrypted to protect privacy.

This subsection examines the ethical and technical challenges of protecting personal data collected by autonomous robots. One of the key challenges is to ensure that data is only collected when necessary and that it is processed in a way that respects the privacy rights of individuals. Techniques such as data minimisation, anonymisation and encryption can be used to reduce the risks associated with data collection and storage, and the potential for misuse of data collected by autonomous

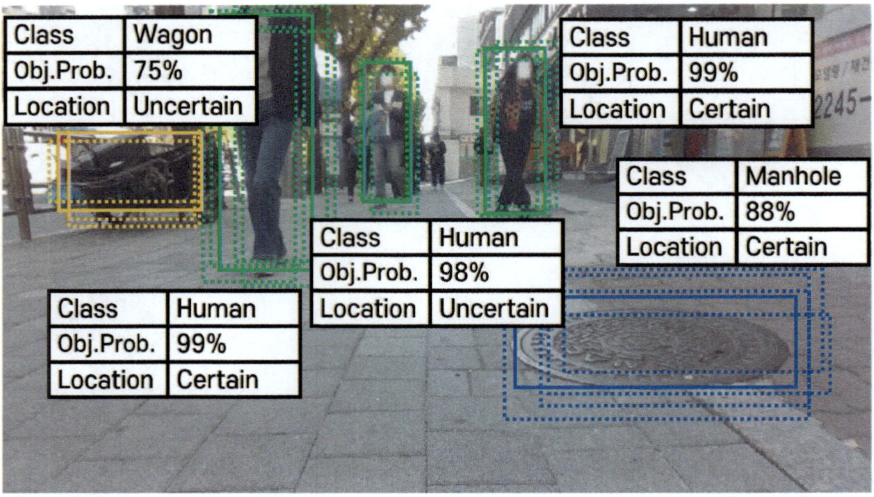

Fig. 7.2 Image collection of autonomous driving car

robots is also a significant concern. Unauthorised access to such data could lead to personal data breaches, surveillance or identity theft. This subsection discusses the importance of establishing strong data protection policies that are in line with existing privacy laws and regulations, such as the European Union's General Data Protection Regulation (GDPR). It also emphasises the role of transparency in data collection practices, which should provide individuals with information about what data is collected, how it is used, and how it is protected (Felzmann et al., 2020). Privacy issues are closely linked to trust and public perception. If people believe that autonomous robots violate their privacy, they are less likely to accept and support the introduction of autonomous robots. Therefore, solving privacy issues is not only an ethical obligation but also an important factor in gaining public trust and promoting the widespread adoption of autonomous robots.

7.1.3 Trust and Public Perception

One of the key challenges of introducing autonomous robots is gaining public trust (Etemad-Sajadi et al., 2022). Trust is a fundamental element of social acceptance, and without it, the deployment of robots will face significant resistance. Public scepticism about autonomous driving technology is growing due to concerns about safety, reliability, employment and potential impacts on privacy. In particular, public concerns about the safety of autonomous driving systems are growing due to a series of accidents involving autonomous driving vehicles that have resulted in injuries or deaths. These accidents are often widely reported in the media, which can create the perception that autonomous robots are inherently unsafe. This subsection examines the factors contributing to public scepticism and discusses strategies for building trust, including transparency in the decision-making process, rigorous safety testing, and public education. Transparency is crucial to building trust. People need to understand how autonomous robots make decisions, how they are tested, and what safety measures are in place to prevent accidents. Providing clear and accessible information on these aspects can help people understand the technology and alleviate fears. Also, involving the public in discussions about the deployment of autonomous robots can foster a sense of shared ownership and responsibility.

This subsection looks at the factors contributing to public scepticism and discusses strategies for building trust, such as transparency in the decision-making process, rigorous safety testing, and public education.

Transparency is crucial to building trust.

People need to understand how autonomous robots make decisions, how they are tested, and what safety measures are in place to prevent accidents. Public demonstrations, pilot programmes and participatory decision-making processes can build trust in the technology by allowing people to see for themselves how it works and how it can benefit their communities. Another important aspect of building trust is ensuring that autonomous robots operate in a way that aligns with social values. This includes not only complying with safety standards, but also ensuring that the robots' actions are perceived as fair and just. For example, if an autonomous delivery robot is programmed to prioritise wealthier neighbourhoods over less affluent ones, it could raise perceptions of bias and inequality, undermining public trust. Therefore, it is essential to design autonomous systems that are not only safe and reliable, but also fair and inclusive, and public education also plays an important role in increasing social acceptance (Naneva et al., 2020). Many people are not familiar with the capabilities and limitations of autonomous robots, which can lead to misunderstanding and fear. The following educational initiatives provide information on how these technologies work and what benefits they offer.

7.2 The Regulatory Framework and Legal Considerations

7.2.1 Development of Regulatory Standards

Clear regulatory standards are essential to ensure the safe deployment of autonomous robots (Yazdanpanah et al., 2023). Regulatory standards provide a framework for assessing the safety, reliability, and ethical behaviour of autonomous systems to ensure that they meet certain requirements before being allowed to operate in public spaces. This subsection looks at the current landscape of regulations related to autonomous systems, including safety standards, testing protocols, and operational guidelines. One of the main challenges in developing regulatory standards for autonomous robots is keeping up with rapid technological advancements. As autonomous technologies evolve, regulations must change to keep pace with new capabilities and potential risks. This requires a dynamic regulatory approach that can respond to new challenges while providing the stability that developers and manufacturers need to innovate with confidence. This subsection discusses the importance of international cooperation to develop harmonised standards that promote the global deployment of autonomous technologies while ensuring consistency of safety and ethical considerations. Regulatory standards should also address the ethical and social implications of autonomous systems, not just the technical aspects. This includes ensuring that autonomous robots operate in a way that is consistent with social values and does not exacerbate existing inequalities. For example, standards could require that decision-making algorithms are free of biases that could lead to discriminatory outcomes.

This subsection discusses the importance of international cooperation to develop harmonised standards that facilitate global deployment of autonomous technologies while ensuring consistency with safety and ethical considerations. Regulatory standards should also address the ethical and social implications of autonomous systems, not just the technical aspects. This includes ensuring that autonomous robots operate in a way that is consistent with social values and does not exacerbate existing inequalities. By including ethical considerations in regulatory standards, policymakers can help ensure that autonomous systems contribute positively to society.

7.2.2 Liability and Legal Responsibility

The introduction of autonomous robots raises complex issues of liability and legal responsibility in the event of accidents or malfunctions. Existing legal frameworks are often not well-suited to addressing the unique challenges posed by autonomous systems, particularly in determining who should be held responsible when problems occur. The issue of liability for accidents is equally imposed on manufacturers, software developers, operators, or robots (Hage, 2017). This subsection examines the existing legal framework and discusses potential approaches to resolving liability issues. One of the key challenges in assigning liability to autonomous robots is the concept of 'autonomous decision-making.' Unlike conventional machines, autonomous robots can make decisions independently based on programming and sensory inputs. This creates a heterogeneity between the actions of human operators and robots, making it difficult to determine who is responsible for those actions. This subsection examines various approaches to liability, including strict liability (where the manufacturer is liable for damages caused by its products) and shared liability models that distribute responsibility among multiple parties, and also discusses the potential need for new legal categories or insurance models tailored specifically for autonomous systems. For example, the concept of 'robot insurance' could be introduced to cover damages caused by autonomous robots, with premiums calculated based on factors such as the robot's intended use, operating environment, and safety record. By developing innovative legal and insurance frameworks, policymakers can mitigate the risks associated with autonomous robots while providing clarity and protection to all stakeholders involved.

7.2.3 *Synergy of Ethics and Regulation*

Effective regulation must balance technological innovation with ethical considerations (Hevelke & Nida-Rümelin, 2015). Autonomous robots have the potential to bring significant benefits to society, but only if they are designed, deployed, and regulated in a manner that is consistent with ethical principles such as fairness, accountability, and transparency. This subsection describes how ethical principles can be incorporated into regulatory frameworks to ensure that autonomous robots operate in a manner that is consistent with societal values. An approach to achieving ethical and regulatory synergies is to develop 'ethical guidelines' that complement existing technical standards. These guidelines can provide a framework for assessing the ethical impact of autonomous systems, including their impact on privacy, equity, and public welfare. By incorporating ethical guidelines into the regulatory process, policymakers can ensure that ethical considerations are not an afterthought in the development and deployment of autonomous driving technology, but rather a core element. The role of regulators in facilitating public engagement and discussion on the ethical implications of autonomous technology is also very important. By involving the public in the regulatory process, policymakers can ensure that the standards and guidelines applied to autonomous robots reflect the concerns and values of a broad community. This subsection emphasises the importance of public consultation, transparency in regulatory decision-making, and the inclusion of diverse perspectives in the development of regulatory frameworks, and goes on to explore the potential of regulatory sandbox controlled environments, where autonomous technologies can be tested and evaluated in real-world conditions without regulatory burdens. Regulatory sandboxes provide valuable insights into the practical challenges and ethical implications of deploying autonomous robots, allowing regulators to improve standards and guidelines based on empirical evidence. In conclusion, developing a regulatory framework and legal considerations for autonomous robots is a complex but essential aspect of their integration into society. Policymakers can support innovation by creating a flexible and adaptive regulatory environment while ensuring that autonomous systems are safe, ethical, and consistent with social values. By emphasising the importance of including diverse perspectives in the development of regulatory decision-making, regulatory sandboxes, and the potential of a controlled environment for testing and evaluating autonomous technologies in real-world conditions without regulatory burdens.

Regulatory sandboxes provide valuable insights into the practical challenges and ethical implications of deploying autonomous robots, allowing regulators to improve standards and guidelines based on empirical evidence. By establishing clear standards, addressing liability issues, and incorporating ethical principles into the regulatory process, policymakers can create an environment in which autonomous robots can thrive while protecting public safety, trust, and well-being.

7.3 The Long-Term Perspective and Social Acceptance

7.3.1 The Role of Public Education and Participation

Active public education and participation are needed to increase social acceptance of autonomous robots. Public participation is not only a means of informing the community, but also an important tool for reducing scepticism and increasing support for the deployment of these technologies. This subsection examines the role of educational programmes, public demonstrations, and participatory decision-making in increasing public understanding and reducing scepticism about autonomous technologies. One of the key challenges to increasing social acceptance is bridging the knowledge gap between autonomous technology developers and the general public. Many people are not familiar with how autonomous robots work, what they can do, and what risks they pose. This lack of understanding can lead to fear, misunderstanding, and resistance to adoption. Educational initiatives can help bridge this gap by providing accessible information about the technology, including how it works, the benefits it provides, and how to manage risks (MacDorman et al., 2009). Public demonstrations and pilot programmes can also play an important role in building public trust and acceptance. Demonstration programmes can help people understand the technology and demonstrate its potential benefits by allowing them to see autonomous robots in action. These programmes can help to alleviate fears and build confidence in safety and reliability by providing the public with the opportunity to experience the technology in a controlled environment. In addition, an interactive decision-making process that involves the community in discussions about where and how to deploy autonomous robots can foster a sense of shared ownership and trust. Public education should also address ethical considerations related to autonomous robots, such as privacy, safety, and fairness. By engaging the public in discussions about these issues, developers and policymakers can ensure that the deployment of autonomous robots aligns with social values and addresses broader community concerns. This proactive approach to public education and engagement is essential to building the trust and acceptance necessary for the successful integration of autonomous robots into society.

7.3.2 Long-Term Vision for Social Integration of Autonomous Robots

The long-term integration of autonomous robots into society requires a clear vision of how these technologies can improve the quality of life and solve pressing social challenges. This subsection provides a forward-looking perspective on the role of autonomous robots in addressing issues such as urban congestion, accessibility, and public welfare. Autonomous robots have the potential to significantly reduce urban congestion by optimising traffic flows, providing efficient last-mile delivery services, and supporting the development of shared mobility solutions. For example, autonomous driving delivery robots can reduce traffic congestion and emissions by reducing the number of delivery vehicles on the road. Similarly, autonomous driving shuttles and ride-sharing services can contribute to more sustainable urban mobility by providing an efficient alternative to private car ownership, and autonomous robots can play an important role in improving accessibility for the disabled and supporting an ageing population, in addition to reducing traffic congestion. Figure 7.3 shows different forms of autonomous delivery robots. Autonomous mobility solutions such as autonomous driving wheelchairs and robot assistants can provide greater independence for people with reduced mobility, improving their quality of life and enabling them to participate more fully in society.

Autonomous robots can contribute to a more inclusive and equitable society by addressing the needs of vulnerable groups, and the long-term vision of autonomous robots also includes the potential to address broader societal challenges such as public health and environmental sustainability. For example, autonomous robots can be used for environmental monitoring, waste management, and pollution control to help achieve urban sustainability goals. In the healthcare sector, autonomous robots can improve the efficiency and effectiveness of healthcare services by supporting tasks such as delivering medical supplies, disinfecting public spaces, and assisting medical staff, but the widespread adoption of autonomous robots also carries potential risks that must be carefully managed. One of the main risks is the potential for jobs to be replaced, especially in sectors such as transportation, logistics, and manufacturing. If autonomous robots replace tasks that have traditionally been performed by humans, there is a risk that workers in those industries could lose their jobs or be forced to transition to new roles. To address this issue, policymakers should develop strategies for retraining and upskilling the workforce to ensure that workers have the skills they need to succeed in an economy increasingly shaped by automation. Another potential risk is that increased reliance on technology could create vulnerabilities in the event of a system failure or cyberattack. Autonomous robots are dependent on a complex network of sensors, data, and software, which can make them vulnerable to technical malfunctions and security breaches. Ensuring the resilience and security of these systems is essential to maintaining public trust and preventing disruption of critical services.

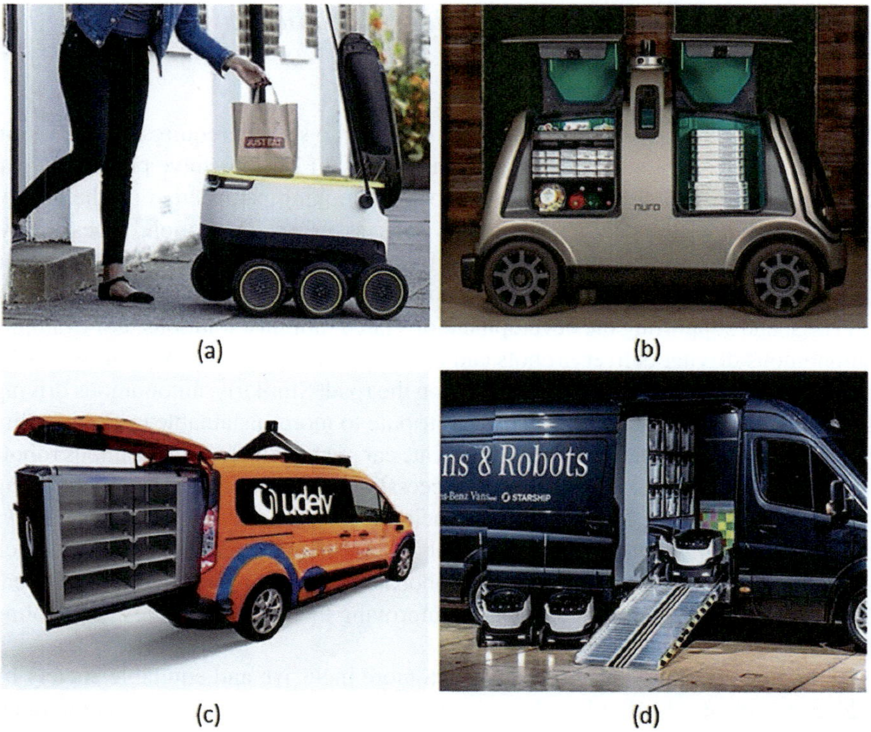

Fig. 7.3 Autonomous driving delivery (Srinivas et al., 2022)

7.3.3 Building a Framework for Social Acceptance

The social acceptability of autonomous robots is important to create an ecosystem that includes regulators, technology developers, and ordinary citizens. This subsection discusses the importance of creating trust through information provision, communicating information in a way that allows non-experts to explain and understand autonomous systems, and the importance of the decision-making process for public feedback.

One of the key elements of social acceptability is making autonomous driving systems explainable (Chatzoglou et al., 2024). People should be able to understand how an autonomous robot makes decisions, what data it uses, and how it interacts with its surroundings. Developing explainable AI (XAI) technology to help people understand the decision-making process of an autonomous robot can help build trust in autonomous robots. By providing a clear explanation of how decisions are made, developers can address concerns about bias and fairness. The decision-making process for public feedback is essential to build trust and ensure that autonomous robots operate in a way that aligns with social values.

In addition, creating a playground can be a way to make autonomous robots more familiar to the public (Choi et al., 2023). In a playground with autonomous robots, children will naturally encounter safety regulations and understand the necessity and convenience. The image of a friendly autonomous robot can improve social acceptance. Figure 7.4 shows an example of a playground with autonomous vehicles.

Public feedback can help developers and policymakers improve their approach to deploying autonomous robots by providing public opinion on community concerns and expectations. Establishing public feedback channels such as community forums, surveys, and participatory workshops allows all stakeholders to be heard and considered in the decision-making process.

Fig. 7.4 PM playground service complex design (Choi et al., 2023)

In addition to information transparency and public feedback, collaboration between regulators, technology developers, and civil society is crucial to creating an ecosystem for the proliferation of autonomous robots. Regulators play a key role in setting standards and guidelines that ensure the safety and ethical behaviour of autonomous systems. Technology developers should work closely with regulators to ensure that innovations comply with these standards and address the concerns of the wider community. Civil society organisations can also play a role in advocating for the interests of marginalised groups and ensuring that the deployment of autonomous robots does not exacerbate existing inequalities. Finally, the social acceptance of autonomous robots depends on their ability to demonstrate their value in solving real-world problems. Developers can help build public support for autonomous robot deployment by focusing on applications that provide real-world benefits to society, such as improving mobility, enhancing public safety, and supporting environmental sustainability. Demonstrating the positive impact of these technologies through pilot projects, case studies and real-world deployments can help build the trust and acceptance necessary for long-term social integration.

In conclusion, the long-term integration of autonomous robots into society requires a comprehensive approach that addresses public education, ethical considerations and regulatory frameworks. Figure 7.5 illustrates the requirements in legal, ethical, and technical domains for enhancing social acceptance. Stakeholders can help ensure that these technologies are accepted by the wider community by promoting public engagement, developing a clear vision of the role of autonomous robots in addressing societal challenges, and building a supportive ecosystem for the deployment of autonomous robots. In this Chapter, we present a roadmap for achieving this vision, emphasising the importance of transparency, inclusivity, and collaboration in successfully integrating autonomous robots into urban environments.

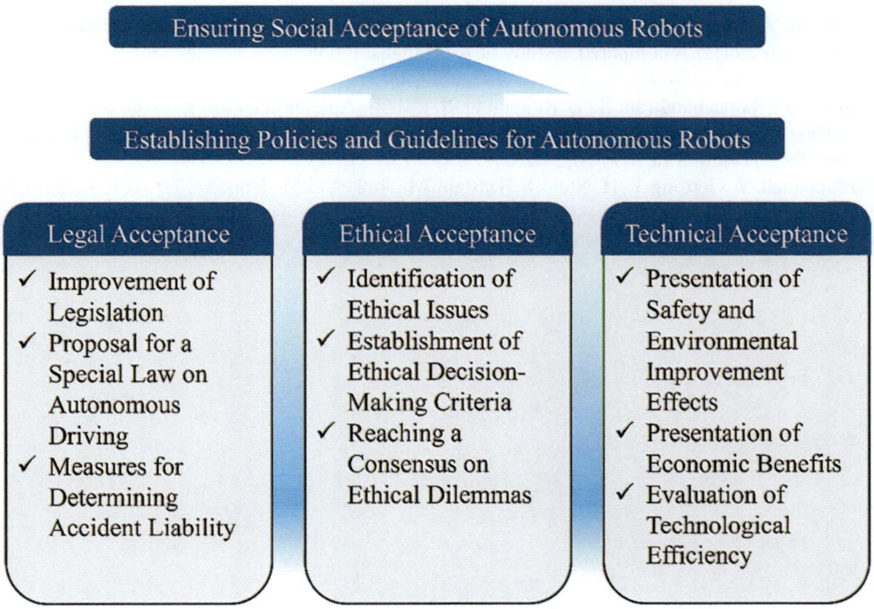

Fig. 7.5 Social consensus for social acceptance

References

Abbass, H. A. (2019). Social integration of artificial intelligence: Functions, automation allocation logic and human-autonomy trust. *Cognitive Computation, 11*(2), 159–171.

Chatzoglou, P. D., Lazaraki, V., Apostolidis, S. D., & Gasteratos, A. C. (2024). Factors affecting acceptance of social robots among prospective users. *International Journal of Social Robotics, 16*(6), 1361–1380.

Choi, M., Ku, D., Kim, S., Kwak, J., Jang, Y., Lee, D., & Lee, S. (2023). Action plans on the reduction of mobility energy consumption based on personal mobility activation. *Energy, 263*, 126019.

Etemad-Sajadi, R., Soussan, A., & Schöpfer, T. (2022). How ethical issues raised by human–robot interaction can impact the intention to use the robot? *International Journal of Social Robotics, 14*(4), 1103–1115.

Farhud, D. D., & Zokaei, S. (2021). Ethical issues of artificial intelligence in medicine and healthcare. *Iranian Journal of Public Health, 50*(11), i.

Felzmann, H., Fosch-Villaronga, E., Lutz, C., & Tamò-Larrieux, A. (2020). Towards transparency by design for artificial intelligence. *Science and Engineering Ethics, 26*(6), 3333–3361.

Hage, J. (2017). Theoretical foundations for the responsibility of autonomous agents. *Artificial Intelligence and Law, 25*, 255–271.

Hevelke, A., & Nida-Rümelin, J. (2015). Responsibility for crashes of autonomous vehicles: An ethical analysis. *Science and Engineering Ethics, 21*, 619–630.

MacDorman, K. F., Vasudevan, S. K., & Ho, C. C. (2009). Does Japan really have robot mania? Comparing attitudes by implicit and explicit measures. *AI & Society, 23*, 485–510.

Naneva, S., Sarda Gou, M., Webb, T. L., & Prescott, T. J. (2020). A systematic review of attitudes, anxiety, acceptance, and trust towards social robots. *International Journal of Social Robotics, 12*(6), 1179–1201.

Pekerti, A. A., & Arli, D. (2017). Do cultural and generational cohorts matter to ideologies and consumer ethics? A comparative study of Australians, Indonesians, and Indonesian migrants in Australia. *Journal of Business Ethics, 143*, 387–404.

Srinivas, S., Ramachandiran, S., & Rajendran, S. (2022). Autonomous robot-driven deliveries: A review of recent developments and future directions. *Transportation Research Part e: Logistics and Transportation Review, 165*, 102834.

Yazdanpanah, V., Gerding, E. H., Stein, S., Dastani, M., Jonker, C. M., Norman, T. J., & Ramchurn, S. D. (2023). Reasoning about responsibility in autonomous systems: Challenges and opportunities. *AI & Society, 38*(4), 1453–1464.

Chapter 8
Environmental Considerations and Sustainability of Autonomous Robots

As the world faces significant challenges due to the climate crisis, international organizations such as the United Nations (UN) are announcing various initiatives and policies to achieve the Sustainable Development Goals (SDGs), which are summarized in Fig. 8.1.

The use of innovative technologies to mitigate the effects of climate change is increasingly emphasized, with autonomous robots playing a crucial role in this discussion. According to the International Federation of Robotics (IFR), robots contribute to achieving 13 out of the 17 United Nations Sustainable Development Goals (SDGs), offering an effective means to conserve resources and develop future green technologies. However, to ensure that robotic advancements align with sustainability objectives, careful strategies must be implemented to minimize their environmental impact and maximize their contribution to ecological conservation.

This chapter focuses on the environmental considerations and sustainability practices involved in the design, deployment, and operation of autonomous robots. A holistic approach is essential, encompassing the entire lifecycle of these technologies—from material selection and energy consumption to end-of-life management. The transformative potential of autonomous robots in urban environments is undeniable. They can enhance quality of life through applications such as goods delivery, personal mobility support, public space maintenance, and air quality monitoring. However, without thoughtful integration and sustainable design, these benefits may come at the cost of exacerbating existing environmental challenges. This chapter explores strategies to ensure the responsible adoption of autonomous robots, concentrating on three key areas: reusable hardware and circular design, carbon neutrality and lifecycle assessment, and the environmental impact of backend support systems.

Sustainability in robotics extends beyond the efficient operation of individual robots; it encompasses the broader ecosystem that supports their functionality. This includes the materials used in robot production, the energy required for manufacturing and operation, and the environmental footprint of the infrastructure supporting

M. Choi and S. Lee, *Intelligent Moving Cities: Technological Leap and Social Integration of Autonomous Mobile Robots*,
https://doi.org/10.1007/978-981-96-8864-7_8

Fig. 8.1 Robots driving sustainability: advancing 13 of 17 UN SDGs through innovation and green technology. *Source* https://ifr.org/ifr-press-releases/news/robots-help-reaching-un-sdgs

data processing and communication. The goal of this chapter is to provide a comprehensive understanding of how autonomous robots can be designed and operated in ways that align with environmental conservation and carbon neutrality objectives.

Section 8.1 explores the principles of reusable hardware and circular design, drawing insights from the aerospace industry, which has successfully implemented reusable components to reduce waste and lower costs. This section highlights the importance of modularity and scalability in robot design, ensuring longer operational lifespans and adaptability to evolving urban needs. By embracing a circular economy model, cities can minimize environmental impact while optimizing resource efficiency in robot production and disposal.

Section 8.2 addresses the necessity of achieving carbon neutrality across the entire lifecycle of autonomous robots. This includes assessing the carbon footprint of manufacturing, identifying strategies to minimize emissions during operation, and integrating renewable energy sources. Lightweight materials, AI-based energy optimization, and lifecycle assessment (LCA) methodologies are key components discussed in this section, providing a structured framework for evaluating and mitigating the environmental impact of robotic systems.

Section 8.3 examines the environmental impact of backend support systems, including data centers and cloud infrastructure, which are integral to autonomous robot operations. These systems require substantial computational power for data processing, decision-making, and communication, often resulting in a significant carbon footprint. This section explores sustainable solutions such as green data centers, energy-efficient edge computing, and the use of renewable energy sources to minimize the environmental burden of these critical support systems.

8.1 Reusable Hardware and Circular Design

Autonomous robots have the potential to transform urban environments, but ensuring that this transformation is environmentally responsible requires a strong focus on the sustainability of robots. This section explores how adopting reusable hardware, modular design, and circular economy principles is crucial in minimizing waste and promoting sustainability.

8.1.1 Inspiration from Aerospace: Lessons from Space-X and Beyond

Space-X, a private space exploration and transportation company founded by Elon Musk, has succeeded in developing reusable rocket technology, significantly reducing the cost and environmental impact of space exploration.

Figure 8.2 shows SpaceX's Falcon 9 rocket launching from ground-based pads, with its Dragon spacecraft separating in the second stage, while the reusable booster returns to Earth for landing. Space-X's Falcon series has revolutionized the economics of space travel and improved sustainability through reusable hardware that can be used multiple times. This concept can be directly applied to the development of autonomous robots. By designing robots that can be reused, repaired, and upgraded—similar to reusable rockets—it is possible to reduce electronic waste and extend the lifespan of robotic systems.

The reuse of robot hardware involves modularization, which allows individual components to be replaced without discarding the entire device. For instance, if a sensor fails or a battery loses efficiency, modular design enables the replacement of just that specific part, thereby minimizing waste. This approach not only helps reduce production costs but also contributes to lowering the overall carbon footprint associated with robot manufacturing.

In the aerospace industry, initial investments in durable and reusable components have been shown to yield significant economic and environmental benefits. According to Harry W. Jones of NASA's research center, the cost of launching to low Earth orbit (LEO) has been reduced by approximately 20 times compared to NASA's traditional space shuttle. He explains that Space-X's reusable launch system has broken historical cost barriers in the space sector. This demonstrates the potential for not only environmental benefits but also economic and industrial achievements through reusable systems (Jones, 2018).

A similar strategy can be employed in autonomous robotics to retrofit and redeploy robots multiple times, aligning with sustainability goals. Focusing on design for reuse is particularly important in urban environments where frequent hardware updates can lead to excessive waste.

Furthermore, leveraging advancements in predictive maintenance inspired by the aerospace sector allows for real-time monitoring of robotic systems to detect wear

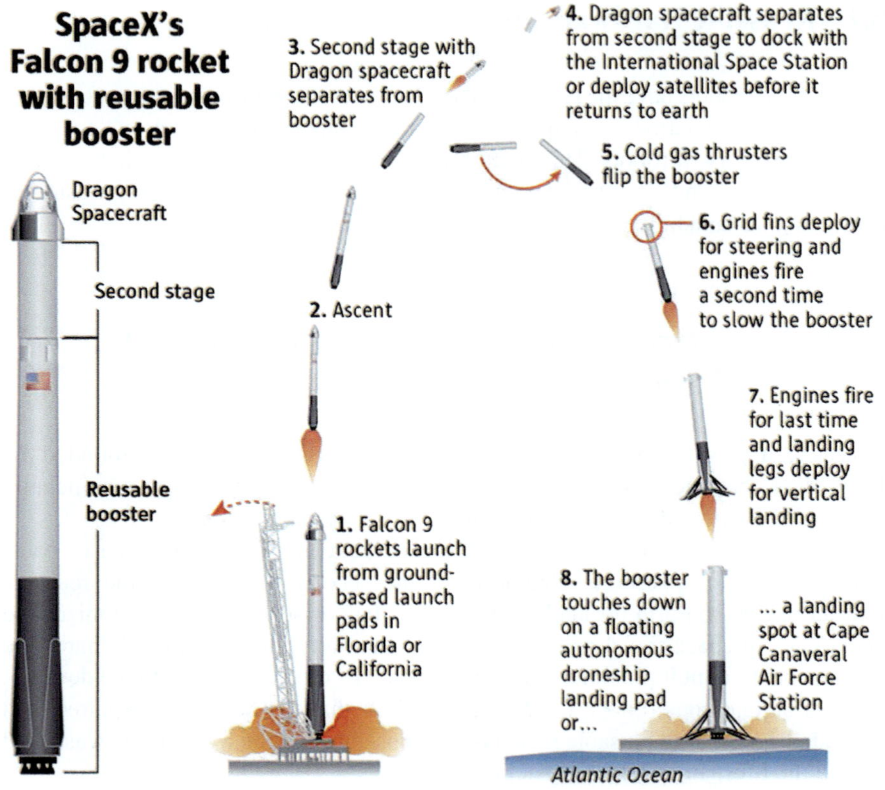

Fig. 8.2 SpaceX's Falcon9 rocket with reusable booster. *Source* https://www.seattletimes.com/business/faa-grounds-spacex-after-rocket-falls-over-in-flames-at-landing/

before failures occur. Predictive maintenance enables repairs or replacements only when necessary, thus extending the robot's lifespan and reducing unnecessary waste. Applying the aerospace industry's emphasis on resilient design to robotic systems can also help them endure harsh urban environments, thereby reducing the frequency of replacements.

8.1.2 Designing Robots with Lifespan and Scalability in Mind: Circular Economy

In urban environments, the design of autonomous robots must consider long-term lifespan, adaptability, and sustainability. Robots must be able to operate for extended periods while withstanding harsh outdoor conditions such as weather, physical impacts, and diverse terrains. To achieve this, durable materials and protective designs should be used to ensure that robots can be used for a long time without frequent replacement. By using environmentally friendly materials like high-strength alloys, composites, and biodegradable materials, the resilience of robots can be enhanced while reducing their environmental impact. This design approach ensures effective operation throughout the robot's lifespan and contributes to minimizing resource wastage.

Japanese researchers have developed a self-growing robot named "GREEMA" for disaster recovery operations using environmentally friendly materials. This robot uses superabsorbent polymers (SAPs) to absorb soil and water in disaster sites, forming a solid structure, which enhances propulsion and mobility (Tsunoda et al., 2023).

As urban environments evolve and grow, scalability is also a key element in robot design. Through modular hardware and software updates, robots can adapt to diverse urban environments. For example, a robot currently supporting pedestrian navigation could expand its role in the future by adding new sensors or tools for cargo delivery. The IT University of Copenhagen in Denmark developed the EMERGE modular robotic platform, which can be assembled in under 30 min using standardized components. The platform utilizes self-connecting modules to quickly change the robot's form and function. In particular, the platform also demonstrated the ability to automatically reposition modules using external robotic manipulators, minimizing human intervention (Moreno & Faiña, 2021).

By designing modular components that can be easily added or replaced, the usability of robots can be increased, and the latest technologies can be continuously integrated to suit various applications. Additionally, by adopting the concept of "design for disassembly," maintenance and upgrades become easier, enabling repairs, component recycling, and repurposing, which helps reduce waste.

The circular economy model is another important element for promoting the sustainability of autonomous robots. In the circular economy approach, robots are not treated as disposable but are designed as systems that can be repaired, modified, and recycled. Figure 8.3 illustrates a general framework of the circular economy, highlighting its key principles. In other words, waste and pollution are designed out, and products and resources are kept in a circular state.

In robot design, it is also necessary to use recyclable, non-toxic materials and adopt a circular economy model where parts from older robots can be reused in new robots or other electronic devices. Thus, the overall flow of robot design and the circular economy is depicted in Fig. 8.4.

However, modularizing each component of a robot or designing it with eco-friendly materials can be technically complex, and therefore, investment in new recycling technologies and processes is also required. Research and development of advanced recycling methods, such as chemical separation technologies or automated disassembly, will play a crucial role in realizing the circular economy for autonomous robots.

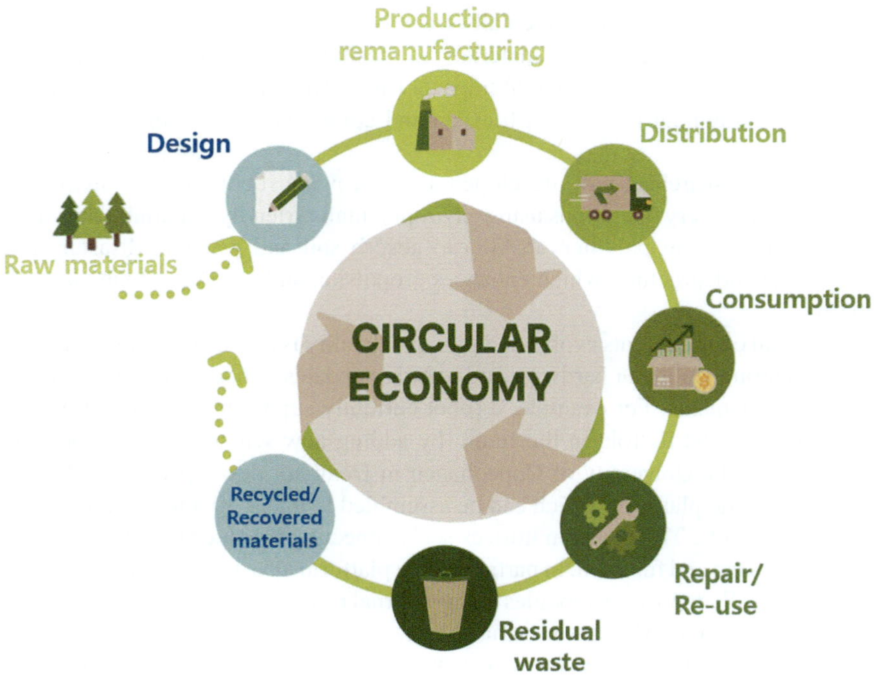

Fig. 8.3 Circular economy: end-of-life products are recycled and recovered instead of becoming residual waste. *Source* https://www.triumvirate.com/blog/do-you-know-about-the-circular-economy

8.2 Carbon Neutrality and Lifecycle Assessment

In the context of autonomous robots, the goal of achieving carbon neutrality is crucial to ensure that the deployment of these technologies does not exacerbate the current climate crisis. This section examines the carbon footprint of robot systems throughout their lifecycle—from manufacturing and operation to disposal—and explores strategies to minimize their environmental impact.

8.2.1 Carbon Footprint of Robot Manufacturing

The manufacturing of autonomous robots involves a series of complex processes, each of which contributes to the overall carbon footprint. Every stage, from raw material extraction to robot component assembly, impacts the environment. This subsection examines the various stages of manufacturing in detail, identifying the major sources of carbon emissions and potential reduction methods.

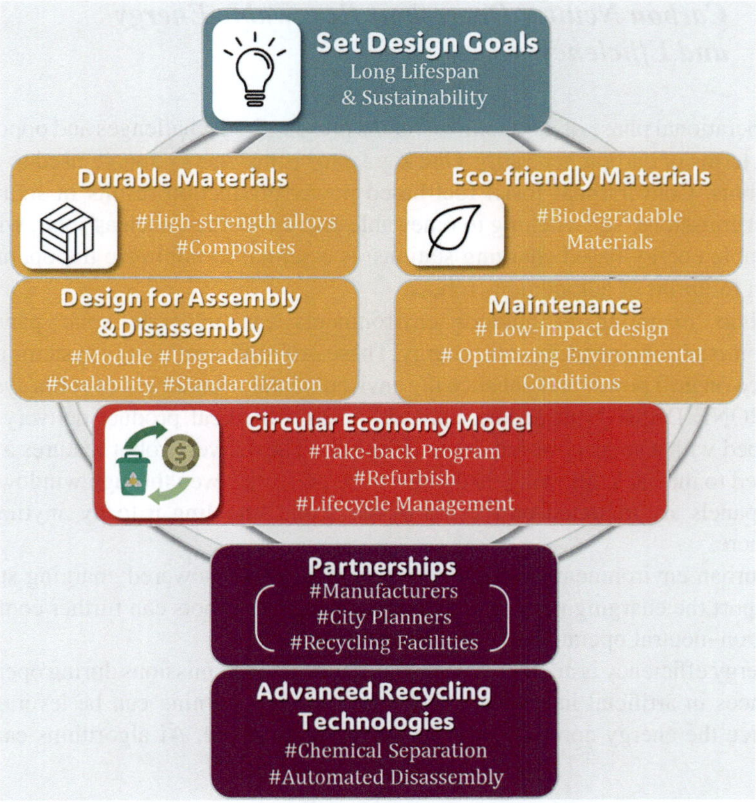

Fig. 8.4 A framework for robot design considering lifespan and sustainability

The production of key robot components, such as batteries, motors, sensors, and control units, generally requires a significant energy input. For instance, the production of lithium-ion batteries, which are commonly used in robots, is associated with high energy consumption and greenhouse gas emissions. To mitigate these impacts, manufacturers can focus on adopting clean energy sources for production facilities, improving energy efficiency, and using recycled or sustainable materials.

Additionally, using additive manufacturing or 3D printing for producing certain robot components has shown potential in reducing waste and energy consumption. Additive manufacturing, which produces only the needed parts, minimizes material waste compared to traditional subtractive methods that cut away excess material. Such changes in manufacturing technologies can help reduce the carbon footprint of autonomous robots.

8.2.2 Carbon Neutral Operation: Renewable Energy and Efficiency

The operational phase of autonomous robots presents both challenges and opportunities in terms of carbon neutrality. One key factor is the energy source used to power the robots. Conventional fossil fuel-based power production results in significant carbon emissions, so switching to renewable energy sources, such as solar, wind, or renewable energy-based charging stations, is essential to minimize the operational carbon footprint of autonomous robots.

Robots operating in outdoor environments can integrate solar panels to autonomously charge using solar energy. These self-sufficient energy systems reduce reliance on grid power and enhance the environmental friendliness of robot systems.

NUONE DESIGN in Milan, Italy, designed a medical product delivery robot equipped with a solar-powered drone (Fig. 8.5). The delivery robot features a drone attached to the back, allowing medicines to be delivered even through windows, and solar panels are mounted on the top of the drone, enabling it to fly anytime and anywhere.

In urban environments, deploying renewable energy-powered charging stations to support the charging needs of multiple autonomous robots can further contribute to carbon–neutral operations.

Energy efficiency is another crucial factor in reducing emissions during operation. Advances in artificial intelligence (AI) and machine learning can be leveraged to optimize the energy consumption of robots. For example, AI algorithms can help

Fig. 8.5 Autonomous robot with solar panels. *Source* https://www.yankodesign.com/2021/08/17/this-medicine-delivery-bot-carries-your-supplies-in-the-last-mile-using-a-solar-powered-drone/

robots plan routes more efficiently, minimizing unnecessary movements and saving energy. Similarly, energy-efficient driving algorithms can be used to reduce power consumption during tasks like navigating urban environments or supporting logistics.

Additionally, using lightweight materials in the design of autonomous robots can reduce energy requirements during operation. The lighter the robot, the less energy it consumes while moving, resulting in significant cumulative savings over extended operation. Therefore, the development of lightweight and durable advanced materials is an important area of research for enhancing the energy efficiency of autonomous robots.

Similarly, in the field of industrial robots, methods to optimize energy consumption have been proposed to reduce operating costs and minimize environmental impact. Research suggests monitoring a robot's energy usage to identify areas for improving energy efficiency and utilizing regenerative braking systems to recover and reuse energy generated during deceleration. It also emphasizes the importance of designing efficient paths and movements for robots to minimize unnecessary movements and reduce energy consumption (Soori et al., 2023).

8.2.3 Lifecycle Assessment for Sustainable Robots

A comprehensive lifecycle assessment (LCA) is essential for evaluating the environmental impact of autonomous robots across their entire lifecycle—from raw material extraction and production to usage, maintenance, and disposal. LCA, as defined by ISO 14040, collects potential environmental impacts throughout the entire lifecycle of a product system and identifies the most impactful stages, providing a foundation for implementing targeted actions to reduce overall emissions and resource use (Fig. 8.6).

The LCA process begins with assessing raw materials, taking into consideration factors such as extraction methods, transportation, and processing. This stage often involves significant energy consumption and environmental degradation, especially when dealing with rare earth metals used in sensors and batteries. By conducting LCA, manufacturers can identify opportunities to replace these materials with more sustainable alternatives or use recycled inputs to reduce environmental impact.

In the operational phase, LCA can be used to quantify energy use and emissions, providing benchmarks for improvement. For instance, manufacturers can compare different autonomous robot models to determine which design features contribute most to energy efficiency and make data-driven decisions to improve sustainability.

The end-of-life stage of the lifecycle is particularly important for ensuring sustainability. Proper disposal, recycling, or repair of robot components can significantly reduce environmental impact. LCA helps manufacturers determine the best strategies for end-of-life management, such as recycling valuable materials or repurposing functional components in new robot systems. Establishing recovery programs is also a key factor in responsibly disposing of robots to minimize waste and reduce landfill burden.

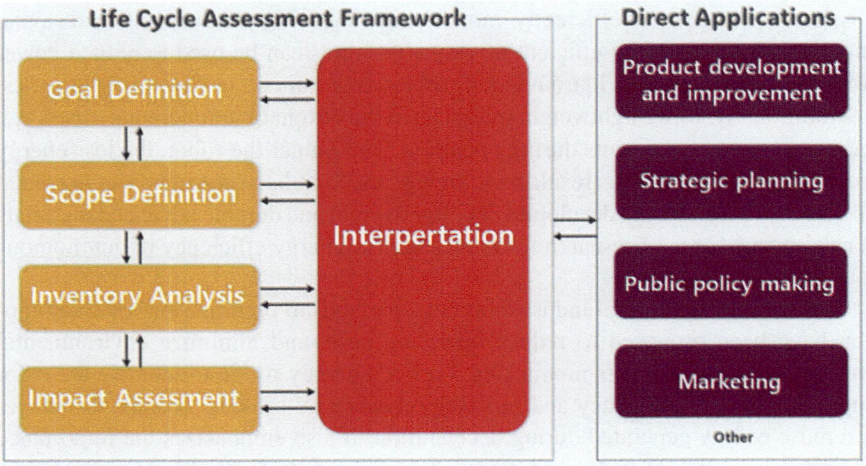

Fig. 8.6 Framework for life cycle assessment (from ISO 14040:2006; modified). *Source* Wolf et al. (2010)

Additionally, LCA data provides insights into the long-term environmental impacts of large-scale deployment of autonomous robots, informing policy decisions. Policymakers can use this information to develop guidelines and incentives that encourage manufacturers to adopt sustainable practices throughout the entire lifecycle of their products.

8.3 Environmental Impact of Backend Support Systems

The environmental impact of autonomous robots extends beyond their physical components to the backend support systems that enable their operation. Systems like data centers and cloud infrastructure have significant environmental footprints. This section examines energy consumption and environmental considerations related to these backend systems and explores strategies to mitigate their impact.

8.3.1 Energy Use in Data Processing and Cloud Support

Autonomous robots heavily rely on backend support systems for data processing, cloud storage, and communication. Data centers, which host servers and provide the computing power needed for robot operation, are energy-intensive facilities. The energy required to process large amounts of data, run complex algorithms, and maintain uninterrupted connectivity significantly contributes to the overall carbon footprint of autonomous robot systems.

According to the International Energy Agency (IEA), there are currently over 8000 data centers worldwide, with approximately 33% located in the United States, 16% in Europe, and nearly 10% in China. In 2022, data centers consumed 460 TWh of energy, and this figure could exceed 1000 TWh by 2026 (Electricity, 2024) (Fig. 8.7).

Specifically, the AI industry is expected to grow exponentially by at least ten times the 2023 demand by 2026, which could lead to an increase in the overall power demand of data centers (Fig. 8.8).

Autonomous robots, which interpret sensor data, make decisions based on AI, and navigate dynamic environments, have high data processing demands. The need for real-time data processing and analysis means that the data centers supporting these robots must operate continuously, leading to significant energy consumption.

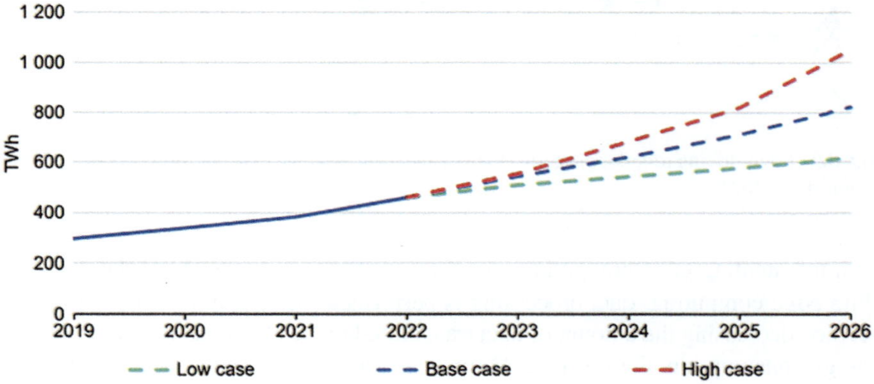

Fig. 8.7 Global electricity demand from data centres, AI, and cryptocurrencies, 2019–2026. *Source* Electricity (2024)

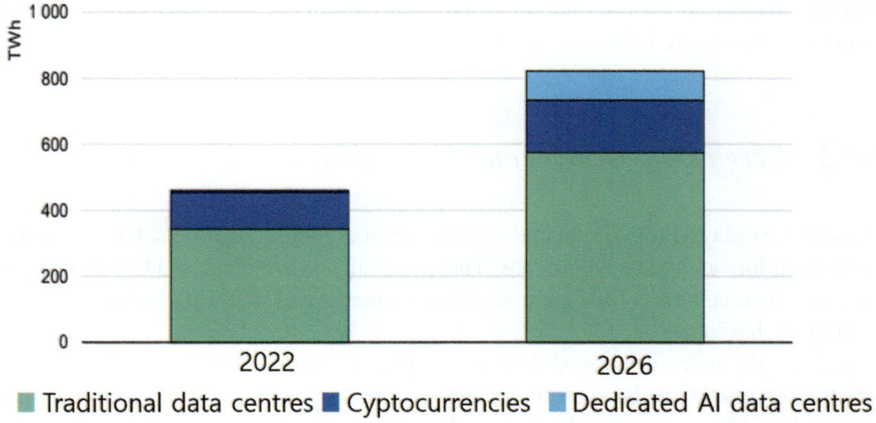

Fig. 8.8 Estimated electricity demand from traditional data centres, dedicated AI data centres and cryptocurrencies, 2022 and 2026, base case. *Source* Electricity (2024)

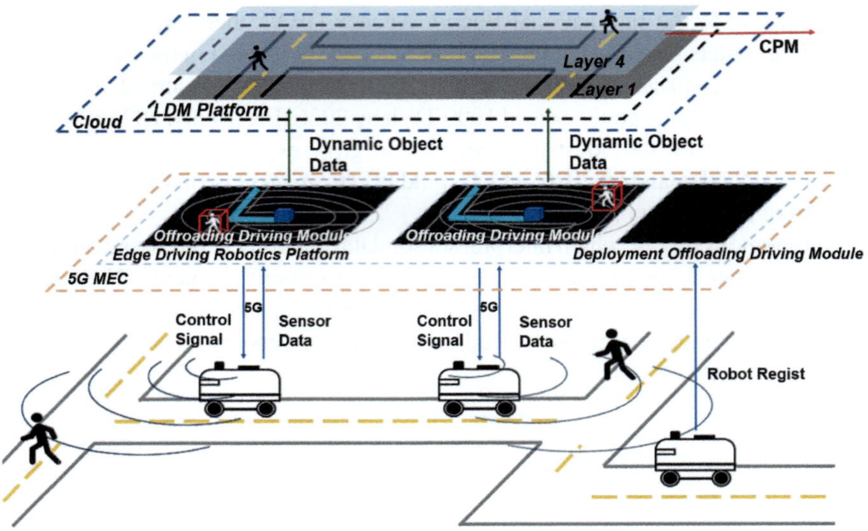

Fig. 8.9 Edge-driving robotics platform (EDRP) and local dynamic map platform (LDMP). *Source* Moon et al. (2024)

Implementing edge computing can reduce reliance on centralized data centers. With edge computing, data processing is performed locally on the robot or nearby servers, decreasing the amount of data transmitted to the cloud and thereby reducing energy consumption. For instance, Moon et al. (2024) proposed an edge computing-based autonomous robot operation platform, demonstrating its potential to lower the power consumption required for robot computing (Fig. 8.9).

The system combines the Edge-Driving Robotics Platform (EDRP) and the Local Dynamic Map Platform (LDMP), and the results demonstrated a 30% reduction in memory usage, an 80% reduction in GPU usage, and a 70% reduction in power consumption required for computation.

8.3.2 Green Data Centers and Sustainable Infrastructure

Data centers play a crucial role in the operation of autonomous robots, but their environmental impact can be significant. Therefore, efforts are being made to transition to green data centers by utilizing renewable energy and adopting energy-efficient cooling technologies.

Firstly, the main drivers of data center power demand are the cooling system and the servers themselves, each typically accounting for 40% of total consumption (Fig. 8.10). Thus, adopting efficient cooling systems has the potential to reduce the power demand of data centers by up to 10% (ABB Motors & Mechanical Inc., 2022). Figure 8.10 shows the energy usage distribution within a data center.

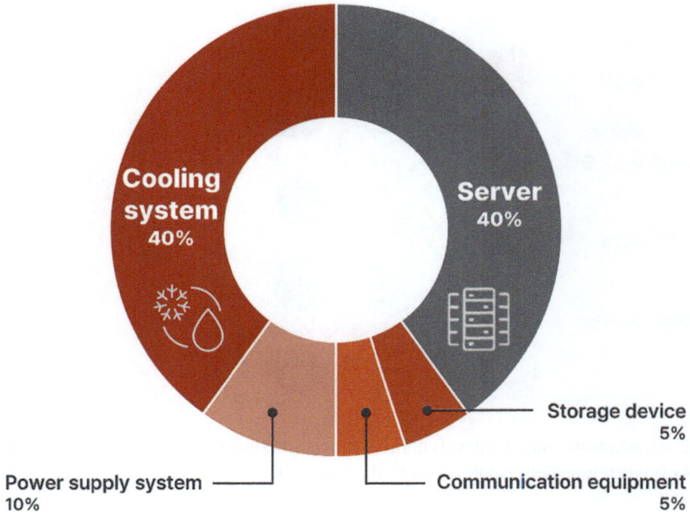

Fig. 8.10 Typical data center energy usage. *Source* ABB Motors and Mechanical Inc. (2022)

Daikin Industries in Japan has developed the Integrated Water-Side Economizer (WSE) technology (i.e., free cooling) for efficient data center cooling systems. This technology is based on the principle that when the external temperature is lower than the cooling system's temperature, the compressor (the main machine used for cooling) does not need to operate, resulting in savings on operational costs and energy. Additionally, the compressor automatically adjusts its compression ratio according to the cooling requirements to maintain optimal efficiency at all times. In other words, energy is used efficiently only when needed, reducing unnecessary power consumption and lowering carbon emissions.

Another key element of sustainable data centers is developing energy-efficient operational measures for data centers. Real-time monitoring and control of data center power consumption can reduce the carbon footprint of data center operations.

For instance, as illustrated in Fig. 8.11, Google has developed and implemented the machine learning-based 'DeepMind AI' framework to understand data center dynamics—such as temperature, power, pump speed, and set points—and optimize Power Usage Effectiveness (PUE).

As a result, they successfully reduced the amount of energy used for cooling by 40%, which is equivalent to reducing the overall PUE overhead energy (including cooling and power conversion) by 15% when considering electrical losses and other non-cooling inefficiencies (Fig. 8.12).

Collaboration among data center operators, robot manufacturers, and urban planners can lead to more sustainable infrastructure. For example, locating data centers near renewable energy sources can reduce transmission losses and enable more efficient use of available energy. Additionally, integrating data centers into urban planning initiatives that prioritize energy efficiency and renewable energy can help create

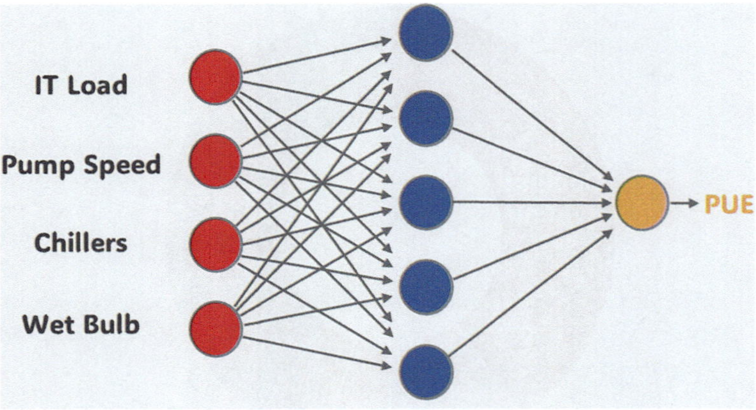

Fig. 8.11 Understanding data center dynamics for optimizing PUE. *Source* https://blog.google/inside-google/infrastructure/better-data-centers-through-machine/

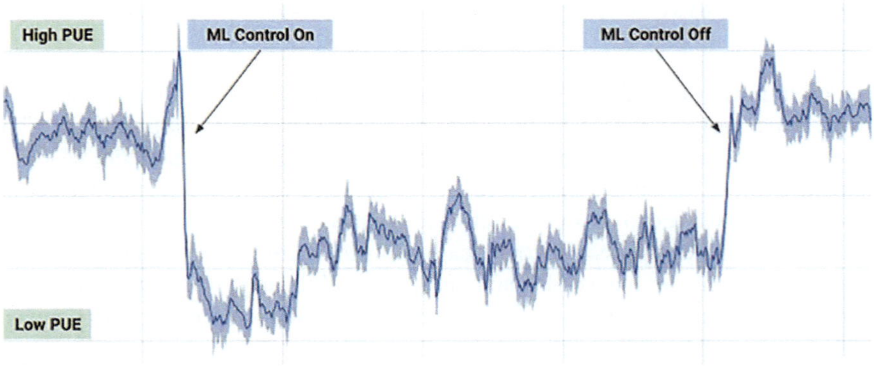

Fig. 8.12 Understanding data center dynamics for optimizing PUE. *Source* https://deepmind.google/discover/blog/deepmind-ai-reduces-google-data-centre-cooling-bill-by-40/

a more sustainable ecosystem to support autonomous robots. In line with this, the revised Energy Efficiency Directive from the European Commission includes regulations that require large data centers to ensure waste heat recovery and utilization (Parlamento Europeo de la 2023).

In the United States, to strengthen power demand management for data centers, the Energy Act of 2020 requires the federal government to conduct research on energy and water use in data centers and produce applicable energy efficiency metrics and best practices to promote efficiency, along with public reporting on past energy and water use in data centers.

The International Telecommunication Union (ITU) developed the framework for reducing energy consumption and carbon emissions for data center owners and developers to mitigate climate change, as detailed in Table 8.1 (McDonald & Ballan, 2023).

Table 8.1 Mitigation strategies for data centers

Sustainable energy	• Plan for renewable energy use, improve energy management, including storage and backup power
Sustainable cooling	• Control temperature • Reduce inefficient cooling and the use of refrigerants • Improve heat and air flow management • Use waste heat • Improve water management, including consumption and sourcing
E-waste management	• Prevent, reduce, reuse, recycle, recover, dispose
Sustainable ICT	• Improve equipment energy efficiency, server use, upgrades and refreshes
Sustainable design and buildings	• Sustainable site selection and use of building materials • Future proofing • Reduce embodied carbon in construction
Implement standards and certifications	• Improve environmental monitoring and reporting

Source McDonald and Ballan (2023)

This framework proposes data center-specific measures to prevent environmental pollution while ensuring data protection.

These efforts combine regulations with practical approaches to enhance the sustainability of data center operations globally.

8.3.3 Autonomous Robots as Environmental Stewards

Autonomous robots have the potential to positively contribute to environmental sustainability by acting as environmental stewards. They can be deployed not only to minimize their own environmental impact but also to carry out tasks that help protect and restore urban environments. This subsection explores how autonomous robots can be used for environmental monitoring, conservation, and public space maintenance.

One of the primary applications of autonomous robots in environmental management is air quality monitoring. Robots equipped with sensors can be deployed across urban areas to collect data on air pollution levels, enabling cities to identify pollution hotspots and take actions to improve air quality. For example, as shown in Fig. 8.13, SMP Robotics' air quality-monitoring robot is equipped with an air monitoring station, allowing it to gather more extensive data compared to stationary stations. Additionally, it has solar panels installed for long-term environmental monitoring.

These robots can contribute to raising public awareness of environmental issues and encouraging more sustainable behavior by providing real-time data.

Fig. 8.13 SMP robotics—air quality monitoring robot (S2.3 Eco). *Source* https://smprobotics.com/products_autonomous_ugv/air-quality-monitoring-robot/

Autonomous robots can also help maintain public spaces by performing tasks such as garbage collection, street cleaning, and waste sorting. Robots equipped with advanced computer vision and manipulation capabilities can identify and collect litter, keeping public areas clean and reducing the burden on human workers.

KOAI Company in South Korea has developed "KOBOT," the world's first autonomous marine pollution control robot. This robot, using autonomous driving technology, selectively collects spilled oil in the ocean and can also collect small solid materials like microplastics (Fig. 8.14). This helps cover areas of the ocean that are difficult for humans to reach, making a significant contribution to addressing marine pollution issues, while also protecting workers' health by handling hazardous chemicals like spilled oil.

Additionally, autonomous robots can be used in environmental conservation efforts, such as reforestation and habitat restoration. Drones and ground-based robots can plant trees in areas affected by deforestation, helping to restore ecosystems and sequester carbon. Moreover, robots can assist in monitoring wildlife, tracking animal populations, and identifying threats to biodiversity, such as illegal logging or poaching.

Fig. 8.14 KOBOT (KOAI)—autonomous marine pollution control robot. *Source* https://koai.co.kr/kobot-remote-control/

Beyond these direct applications, autonomous robots can contribute to sustainability by supporting the implementation of eco-friendly infrastructure. For instance, robots can assist in the construction and maintenance of green roofs, vertical gardens, and urban farms, helping to reduce urban heat islands, improve air quality, and increase urban biodiversity. By automating labor-intensive tasks, robots can enhance the feasibility and cost efficiency of developing green infrastructure, encouraging wider adoption.

References

ABB Motors and Mechanical Inc. (2022). *Motors in data centers powering the connected world*

Electricity IEA. (2024). Analysis and forecast to 2026. In *IEA Report*. https://www.iea.org/reports/electricity-2024

Jones, H. W. (2018). *The recent large reduction in space launch cost*

McDonald, R., & Ballan, S. (2023). *Green data centers: Towards a sustainable digital transformation—A practitioner's guide*

Moon, J., Hong, D., Kim, J., et al. (2024). Enhancing autonomous driving robot systems with edge computing and LDM platforms. *Electronics (Basel), 13*, 2740.

Moreno, R., & Faiña, A. (2021). EMERGE modular robot: a tool for fast deployment of evolved robots. *Front Robot AI, 8*,. https://doi.org/10.3389/frobt.2021.699814

Parlamento Europeo de la UE. (2023). Directive (EU) 2023/1791 of the European Parliament and of the Council of 13 September 2023 on Energy Efficiency and Amending Regulation (EU) 2023/955 (Recast) (Text with EEA Relevance). *Official Journal of the European Union, 50*, 1–111.

Soori, M., Arezoo, B., & Dastres, R. (2023). Optimization of energy consumption in industrial robots, a review. *Cognitive Robotics, 3*, 142–157.

Tsunoda, Y., Sato, Y., & Osuka, K. (2023). *GREEMA: proposal and experimental verification of growing robot by eating environmental material for landslide disaster*. https://doi.org/10.20965/jrm.2024.p0415

Wolf, M.-A., Chomkhamsri, K., Brandao, M., et al. (2010). *International Reference Life Cycle Data System (ILCD) handbook-general guide for life cycle assessment-detailed guidance*

Appendix
Automated Pedestrian Environment Assessment Using Autonomous Robots

A.1 Overview of Automated Pedestrian Environment Assessment

In pedestrian environment assessment plans, sidewalks are often assessed based on subjective criteria. However, when subjective assessments are involved, the results lack consistency and fail to ensure objectivity. Without an objective assessment of sidewalks, areas that require improvement may be neglected, while unnecessary improvements could lead to wasted budgets and various other issues.

Currently, sidewalk assessment systems rely on manual labor and subjective judgments, resulting in inconsistencies across different regions. This highlights the growing need for a standardized pedestrian environment assessment system.

As urban environments shift from being vehicle-centered to pedestrian-centered, the importance of assessing sidewalks has increased. However, there is a lack of pedestrian solutions aimed at improving accessibility and addressing potential hazards. To accurately estimate pedestrian travel conditions, it is essential to distinguish between different modes of movement. When a problem arises on the sidewalk, recognizing and resolving the issue can take a considerable amount of time. Moreover, a large workforce is required for sidewalk assessment, and since human judgment plays a role in the assessment process, consistency and reliability are compromised.

To create a more comfortable and efficient pedestrian environment, this study proposes an assessment method that utilizes deep learning-based image processing algorithms to analyze sidewalk conditions. By defining specific criteria for sidewalk conditions and classifying sidewalk images accordingly, the goal is to establish a systematic approach to automated pedestrian environment assessment.

M. Choi and S. Lee, *Intelligent Moving Cities: Technological Leap and Social Integration of Autonomous Mobile Robots*, https://doi.org/10.1007/978-981-96-8864-7

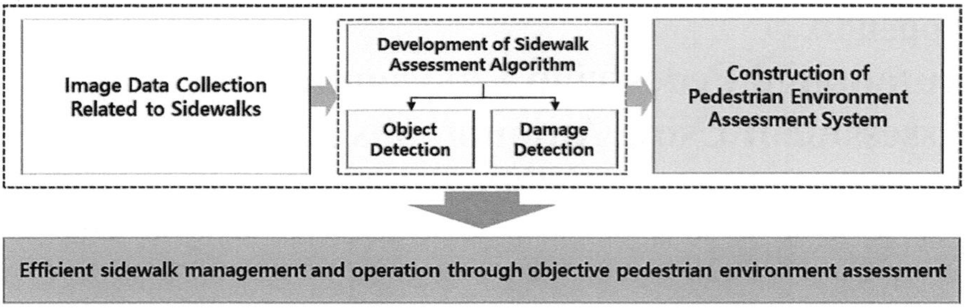

Fig. A.1 Autonomous robot-based framework for pedestrian environment assessment

A.2 The Need for Automated Pedestrian Environment Assessment

If artificial intelligence can autonomously analyze images using deep learning and produce consistent results, it will enhance the objectivity of the assessment.

To develop an automated sidewalk assessment system by building a comprehensive sidewalk image dataset, the following steps are necessary. First, it is essential to investigate the characteristics of obstacles commonly found on sidewalks and establish a survey plan. Based on this plan, image data collection is conducted. Next, a deep learning strategy is designed to classify sidewalk obstacles and detect surface damage. Subsequently, an algorithm is developed using the collected images to assess sidewalk conditions. Once different algorithms have been developed based on specific characteristics, they must undergo rigorous validation to ensure their applicability in real-world scenarios. Ultimately, the goal is to develop an automated sidewalk assessment system based on these validated assessment algorithms. The sidewalk assessment results generated by this automated system can be used to establish priorities for efficient sidewalk management and maintenance. The overall process of this study is illustrated in Fig. A.1. For the implementation of autonomous robots on sidewalks, a variety of sidewalk-related datasets are required for comprehensive assessment and detection. Compared to roadways, sidewalks experience a higher frequency of real-time events, yet there is a lack of established datasets. Currently, no dedicated sidewalk dataset exists that distinguishes between fixed obstacles (which cannot be improved) and non-fixed obstacles (which can be mitigated). To facilitate the implementation of autonomous robots on sidewalks, this study aims to develop a structured dataset that categorizes obstacles into fixed and non-fixed types, as depicted in Fig. A.2.

Through image deep learning, it has been confirmed that detecting sidewalk obstacles and identifying sidewalk damage require different algorithms, despite their shared objective of improving pedestrian safety and providing information for autonomous navigation systems. This distinction primarily arises due to differences in the data types involved, the unique characteristics that need to be recognized, and the varying impacts of these tasks in real-world environments. Sidewalk obstacles

Fig. A.2 Pedestrian environment assessment factors

can include various objects such as trees, streetlights, parked vehicles, and pedestrians. These obstacles are generally three-dimensional and vary significantly in size, shape, color, and texture.

In contrast, sidewalk damage, such as cracks, potholes, and uneven surfaces, is typically two-dimensional or shallow in depth. Therefore, it requires a focus on texture and surface patterns rather than object shape. Due to the distinct characteristics of these two types of data, different algorithmic approaches are required for efficient and accurate detection. To effectively identify obstacles, the algorithm must be trained to recognize a wide range of object classes. On the other hand, damage detection requires the ability to identify anomalies on an otherwise uniform surface. This necessitates algorithms capable of detecting pattern deviations, texture changes, or disruptions in continuity. The actions required after detecting obstacles versus identifying damage also differ. For instance, real-time detection and response are crucial for obstacles, as they need to be navigated around. In contrast, damage detection does not necessarily require an immediate response but is critical for maintenance planning and ensuring long-term pedestrian safety. These differing requirements influence the most suitable algorithm types for each task.

Object detection algorithms commonly used for obstacle detection are designed to identify and classify multiple objects within a given image. Algorithms such as YOLO (You Only Look Once) and SSD (Single Shot MultiBox Detector) are capable of handling various object classes, and this study specifically focuses on YOLO for analysis. Conversely, for damage detection, CNN (Convolutional Neural Networks) are commonly employed, as they are designed to detect deviations from normal patterns, making them more suitable for identifying cracks, potholes, and other forms of sidewalk damage. In this study, the CNN architecture will be modified to enhance fine-grained detection and compared against existing algorithms. In conclusion, while both tasks aim to enhance sidewalk safety, the differences in data types, functional objectives, real-world impact, and optimal algorithmic approaches necessitate the development of separate algorithms for sidewalk obstacle detection and damage identification.

A.3 Application of Autonomous Robots in Pedestrian Environment Assessment

The collected sidewalk images were categorized into 12 distinct classes, allowing for precise analysis and classification. Based on this, algorithms were developed to identify sidewalk damage, assess the severity of the damage, and integrate these functions into a comprehensive system for evaluating sidewalks. Recently, autonomous robots on sidewalks have been gaining traction, and various types of autonomous sidewalk robots are equipped with sensors such as cameras and LiDAR to collect sidewalk images. By applying the developed algorithms to the collected images, the pedestrian environment assessment system can be continuously updated and improved. This data can contribute to creating an optimized environment for seamless movement on sidewalks for autonomous robots and platforms, ultimately enhancing navigation efficiency. Therefore, algorithms for detecting obstacles and sidewalk damage have been applied to the sidewalk robot system. The functionality of these algorithms was verified through scenario-based analysis, ensuring that they operate smoothly from the robot's perspective and movements. As the world becomes increasingly automated, the application of robotics in various fields is becoming more common. One critical challenge faced by autonomous robots is their ability to effectively navigate their surroundings, detect obstacles in their path, and avoid them.

To address this challenge, an autonomous robot system was assembled, equipped with high-resolution cameras and LiDAR sensors to perceive and analyze the surrounding environment. At the same time, a deep learning model was developed to analyze images and detect potential obstacles.

By combining the sidewalk obstacle detection algorithm with the autonomous robot system, we aim to identify and collect data on obstacles, and upon detection, enable the robot to avoid them. This system will be integrated into the existing sidewalk assessment model to determine the feasibility of autonomous navigation on sidewalks.

A.4 Test Autonomous Robot and Hardware Specifications

For the experiment, the WeGO ST (Scout)-Mini, an autonomous-capable robot, was used. Figure A.3 shows the WeGO ST-Mini with sensors mounted, while Fig. A.4 presents the design of the ST-Mini without sensor mounting. Table A.1 summarizes the key specifications of the ST-Mini. The ST-Mini is a compact four-wheel differential drive platform with independent suspension, capable of reaching a maximum speed of 10 km/h. It serves as a UGV (Unmanned Ground Vehicle) platform for research and educational purposes in autonomous driving. It supports RS-232 communication and allows flexible programming with preferred software,

Table A.1 Specifications and features of the RealSense camera

Hardware	Specification
Size	627 × 550 × 252 mm
Mode of operation	4 wheel drive, differential—drive model
Wheelbase	452 mm
Battery operation temperature	− 20 to 60°
Minimum ground clearance	107 mm
Minimum turning radius	0 m
Max velocity	10 km/h
Communication environment	Standard CAN/RS232

Fig. A.3 WeGo ST-mini

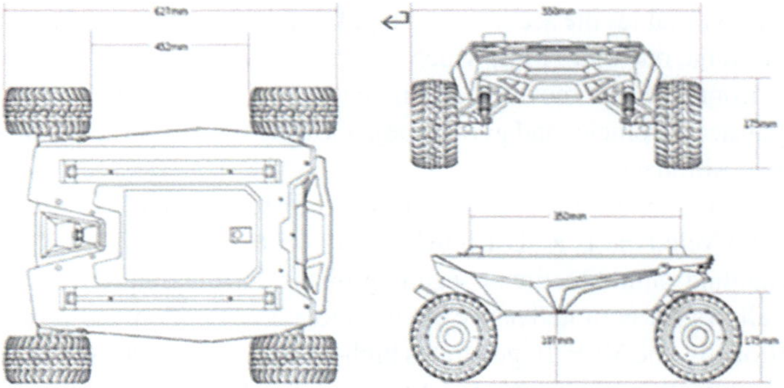

Fig. A.4 Size of robot

making it a versatile autonomous driving research and development platform. Additionally, the ST-Mini features four-wheel drive capability and a top-mounted profile structure, which facilitates easy sensor mounting.

The test robot, WeGO ST-Mini, is equipped with various mounted sensors. It includes a RealSense camera, which can simultaneously capture RGB images and

Table A.2 Specifications and features of the RealSense camera

Sensor	RealSense depth camera D435i
Manufacturer	Intel
Features	Captures depth and color images, depth sensing, pose tracking, high-speed data capture, supports various software
Resolution	1920 × 1080 (color), 1280 × 720 (depth)
Maximum measurement distance	10 m
Frame rate	Up to 90 FPS
Sensor type	Stereo camera, depth sensor, IMU
Depth resolution	1280 × 720
Depth accuracy	± 2% (within 0.2–1.5 m range)
IMU	6-axis IMU (accelerometer, gyroscope)
IMU update rate	Up to 400 Hz
Supported software	Intel RealSense SDK, ROS, etc.

depth information, and a Velodyne LiDAR, capable of collecting 3D data. Additionally, it is equipped with an RTK GPS, which provides high accuracy and reliability, and an IMU (Inertial Measurement Unit) for measuring the position, velocity, and orientation of objects.

For recognizing and identifying images of sidewalk obstacles and surface damage, the RealSense D435i camera was used. Table A.2 summarizes the specifications and features of the D435i. The D435i camera is utilized for environment perception, as it can capture both depth and color images. With depth sensing capabilities, it can measure object distances and assist in obstacle detection. Its high-speed data capture function enables the accurate and rapid acquisition of image data, supporting a maximum frame rate of 90 FPS, which ensures precise image capture even in fast-moving environments. Moreover, the camera is equipped with a built-in 6-axis IMU (including an accelerometer and gyroscope), allowing for the extraction of additional motion and positional data.

Table A.3 summarizes the specifications and features of the VLP-16. The VLP-16 LiDAR by Velodyne is a 3D LiDAR sensor with a 360-degree horizontal field of view, enabling omnidirectional scanning of the surrounding environment. This capability allows for comprehensive environmental perception and obstacle detection. Additionally, the VLP-16 generates high-density point clouds, which facilitate accurate identification of obstacle positions and distances. This information can be utilized for robot path planning, collision avoidance, and other safety functions. The VLP-16 is equipped with 16 channels, supports a maximum measurement range of up to 100 m, and features an adjustable rotation speed ranging from 5 to 20 Hz.

Table A.4 summarizes the specifications and features of the MRP 2000. The MRP 2000 RTK-GPS by Synerex is equipped with multi-GNSS (Global Navigation Satellite System) data processing capabilities, enabling accurate position measurement

Table A.3 Specifications and features of VLP-16

Sensor	VLP-16
Manufacturer	Velodyne
Features	360° environment scanning, obstacle detection, terrain analysis, multi-return mode, supports both indoor and outdoor environments
Channels	16 channels
Measurement Range	100 m
Accuracy	Up to ± 3 cm(typical)
Vertical field of view	+ 15.0° to − 15.0°
Horizontal field of view	360°
Vertical angular resolution	2.0°
Horizontal/azimuth angular resolution	0.1°–0.4°
Rotation speed	5–20 Hz

Table A.4 Specifications and features of MRP 2000

Sensor	MRP 2000
Manufacturer	Synerex
Features	Provides real-time accurate position measurement and precise position estimation capability
Operating Frequency	L1/L2 broadcast RTK data reception capability
RTK accuracy	10 mm + 1 ppm CEP
RTK fix time	10–60 s
Power input	DC 5 V/2 A (microUSB)
Dimensions	46.5 × 97.4 × 24 mm (W × H × D)
Weight	98 g
Operating temperature	− 25 to 85 °C

and estimation. Additionally, it receives L1/L2 broadcast RTK data, providing real-time, highly accurate positioning information. The MRP 2000 offers centimeter-level precision, with a positioning accuracy of 10 mm + 1 ppm CEP and a fix time ranging between 10 and 60 s. This ensures rapid and precise position measurement support. By leveraging these capabilities, MRP 2000 is used to obtain GPS-based distance data for robot navigation, perform accurate position measurements, and enhance position estimation.

Table A.5 summarizes the specifications and features of the VN100. The VN100 IMU (Inertial Measurement Unit) by VectorNav provides precise position estimation based on an accelerometer and gyroscope. By measuring the rotation speed from the gyroscope and the acceleration from the accelerometer, it enables accurate estimation of a robot's or vehicle's position and orientation. The VN100 consists of a 3-axis accelerometer and a 3-axis gyroscope, allowing multi-axis measurement for highly

Table A.5 Specifications and features of VN100

Sensor	VN100
Manufacturer	VectorNav
Features	Accurate attitude estimation and motion detection, multi-axis measurement, velocity data extraction
Accelerometer range	$\pm$ 16 g
Gyroscope range	$\pm$ 2000°/s
Extended Kalman Filter update rate	Up to 400 Hz
IMU data rate	800 Hz
Heading	2.0°
Pitch/roll	0.5°
Power	220 mW
Linear bias acceleration	< 0.04 mg
Gyroscope in-run bias stability	5–7°/hr

accurate tracking of movement and posture across different axes. Additionally, it supports high update rates, ensuring rapid data acquisition. Due to these capabilities, the VN100 is widely used in various fields, including robotics, aerospace, automotive, and autonomous driving applications.

Figure A.5 presents the topic information of WeGO ST-Mini before conducting the experiment. This provides insights into the communication status between nodes using ROS (Robot Operating System). Table A.6 is a summary table of the topics for publishers and subscribers. A publisher acts as the sender that publishes messages, while a subscriber acts as the receiver that receives messages. The publisher publishes the messages Scout_Status, /odom, and /tf. Scout_Status provides status information of the Scout Mini, /odom transmits the odometry information of the robot, and /tf provides coordinate transformation information of the robot. The subscriber receives the data from /scout_light_control and /cmd_vel. /scout_light_control receives messages related to the robot's light control, and /cmd_vel receives movement commands for the robot.

The camera recognizes images in front by utilizing light and can analyze incoming images at the pixel level or based on image shapes to make judgments. Table A.7 summarizes the role of cameras in robots, demonstrating that they can perform a variety of functions beyond simply capturing images. The roles of a camera mounted on a robot include vision-based navigation, localization and posture estimation, image-based detection and analysis, environmental modeling and mapping, and vision-based operations. Figure A.6 shows various types of cameras used in robots.

LiDAR is a sensor that uses laser signals to recognize surrounding objects. When the emitted laser signal hits an object and reflects back, the sensor analyzes it to determine the object's position, movement direction, and speed. Unlike cameras,

Table A.6 ROS topic

ROS TOPIC		
Publisher	/scout_status	Outputs the actual status of the Scout Mini
	/odom	Check the current position and movement status based on wheel displacement
	/tf	Provide transformation values between ROS frames
Subscriber	/scout_light_control	Enable front light control of the Scout in autonomous mode
	/cmd_vel	Control speed and rotational velocity in autonomous mode

Fig. A.5 Scout mini ROS Topic

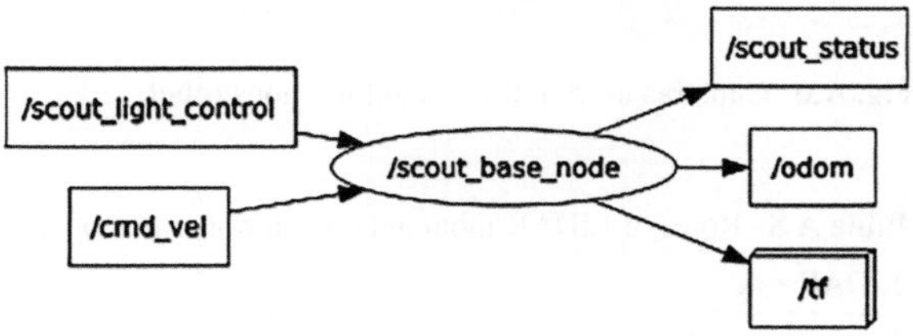

Table A.7 Roles of cameras mounted on autonomous robots

Camera role	
Vision-based exploration	Recognizes the surrounding environment, detects obstacles, and assists in finding targets, enabling the robot to move autonomously and adapt to its surroundings
Localization and posture estimation	Determines the robot's position and estimates its orientation in space, allowing it to recognize objects, identify their location, and move accordingly
Image-based detection and analysis	Uses image data to detect various features such as shape, color, and objects, and provides environmental information
Environmental modeling and mapping	Assists in modeling the surrounding environment Identifies objects and detects defects in products Performs visual tasks based on images or drawings
Vision-based task execution	Supports the execution of vision-based tasks Varies depending on the robot's application and purpose, and enhances the robot's perception and functionality through integration with various sensors

LiDAR in robots has the capability to detect objects in front of it and change direction accordingly. Table A.8 summarizes the roles of LiDAR in robots, which include distance and depth measurement, environmental perception and obstacle detection,

Fig. A.6 Cameras used in the test autonomous robot

Table A.8 Roles of LiDAR mounted on autonomous robots

LiDAR role	
Distance and depth measurement	Accurately measures the distance and depth of the surrounding environment Detects the distance of obstacles around the robot and uses this information for safe path planning
Environmental perception and obstacle detection	Recognizes the surrounding environment Detect obstacles such as walls, vehicles, and other objects to avoid collisions or navigate safely
Mapping and robot localization	Can perform 3D mapping of the surrounding environment and estimate its own position Constructs a map of surrounding structures and obstacles, performs real-time position estimation, and maintains accurate posture and location
Autonomous navigation and path planning	Recognizes terrain and obstacles in the environment and plans a safe path for movement Avoids traffic obstacles or adjusts driving speed to enable efficient autonomous navigation
Object recognition and tracking	Analyzes LiDAR data in 3D space to identify and track objects Detect moving objects and track them for interaction and response

mapping and robot localization, autonomous navigation and path planning, as well as object recognition and tracking. Figure A.7 illustrates the different types of 2D and 3D LiDAR used in robots, which vary according to the robot's application and purpose.

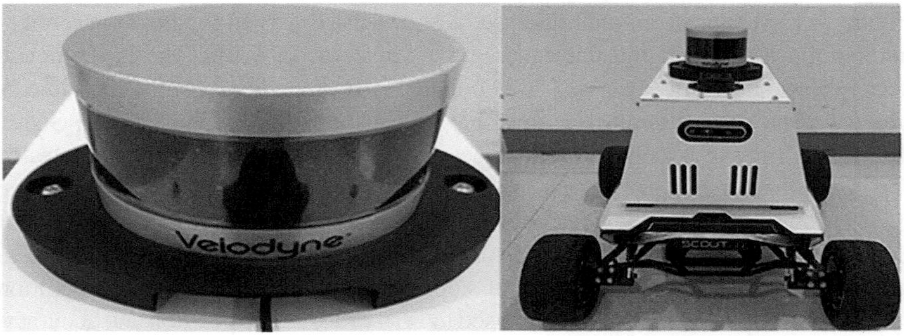

Fig. A.7 LiDAR used in the test autonomous robot

A.5 Pre-driving Simulation

Before conducting the experiment, a simulation was performed to analyze how the robot recognizes obstacles and determines its driving path for avoidance. The simulation was conducted using ROS (Robot Operating System) within the Gazebo environment. Figure A.8 shows an image with adjusted camera parameters. While camera calibration is not required in the simulation, it may be necessary during actual robot testing. Camera calibration is performed to enhance the clarity of detected obstacles or lane markings in the camera's field of view. Figure A.9 illustrates the process of modifying the image perspective by adjusting parameter values using a red trapezoid. If the camera receives unclear images, parameter adjustments can be made to enhance image clarity and optimize projection settings.

The simulation tested how the robot determines its driving path and avoids obstacles when rectangular obstacles are placed in its way. Figure A.10 presents a simulation where an obstacle is positioned in front of the robot, and the robot is given a driving command to detect and avoid the obstacle. The simulation results confirmed that the robot avoided all three obstacles without collision. The robot detected the obstacles through its camera, stopped, modified its driving path, and then resumed movement to reach its destination. Through this simulation, the robot's obstacle avoidance strategy and driving path were analyzed. In pedestrian environment testing, the trained obstacles were used to ensure that the robot could detect, assess, stop, and navigate around obstacles. Additionally, various simulations involving traffic signs, traffic signals, crosswalks, and tunnels have been included in the appendix.

A.6 Validation Scenario Configuration

This study aims to develop a technology that enables a robot to identify and assess obstacles on sidewalks using images collected by a RealSense Depth Camera (D435i) mounted on the robot, and to apply an obstacle avoidance algorithm to the robot.

Figure A.11 shows the simulation-based test process designed to allow the robot to autonomously navigate the sidewalks of the University of Seoul. The experimental scenario, as shown in Fig. A.12, was conducted by selecting areas with sidewalk obstacles and surface damage within sidewalk sections of the University of Seoul. A total of three test sections were selected. In each section, the robot was tested for its ability to recognize, assess, and avoid various sidewalk obstacles, including personal mobility (PM) devices, bicycles, pedestrians, and pavement damage. For this experiment, the WeGO ST-Mini robot was used. The RealSense Depth Camera was employed to collect image data of sidewalk obstacles and train the recognition system. Additionally, RTK-GPS was used to verify the robot's position and accuracy, and the experiment was conducted in a ROS (Robot Operating System) environment to send avoidance and movement commands to the robot and apply the algorithm. Through this study, the performance of the robot in recognizing, assessing, and safely avoiding sidewalk obstacles is evaluated. The findings aim to contribute to enhancing the stability and efficiency of autonomous robots.

The first test section was a 50-m stretch of Sidewalk 1 at the University of Seoul, as shown in Fig. A.13. In this section, PM (Personal Mobility) devices and surface damage were tested as sidewalk obstacles. For the experiment, a starting point and end point were designated, with a waypoint set along the path. To verify whether the robot could accurately detect and avoid obstacles in this section, GPS data was collected in advance using RTK-GPS. During the test, the algorithm developed in the ROS environment and pre-trained data on sidewalk obstacles were applied to the robot. The test assessed whether the robot could detect PM devices and damage in

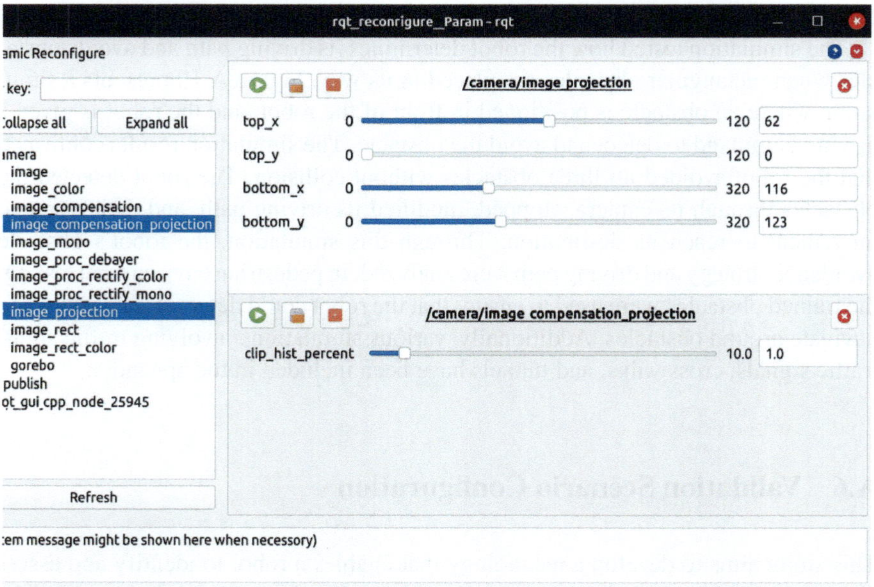

Fig. A.8 Camera parameter adjustment—rqt_reconfigure_param

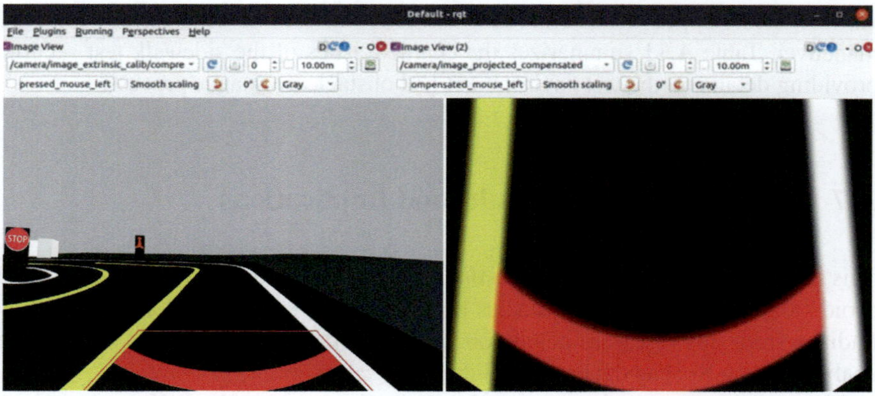

Fig. A.9 Camera and ground projection image after camera parameter adjustment (simulation)

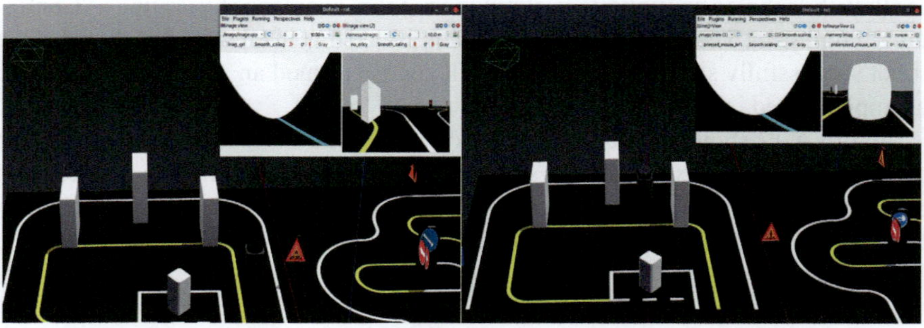

Fig. A.10 Camera and ground projection image after camera parameter adjustment (simulation)

the test section, stop, and then perform an avoidance maneuver after moving a certain distance. Table A.9 summarizes the characteristics of Sidewalk 1 at the University of Seoul, providing information on the length of the section, types of obstacles, driving method, and avoidance points.

The second test section was a 45-m stretch of Sidewalk 2 at the University of Seoul, as shown in Fig. A.14. In this section, pedestrians were selected as the primary obstacle for testing. For the experiment, a starting point and an end point were designated, with a waypoint set along the path. The test was conducted to verify whether the robot, upon detecting a pedestrian from its learned image dataset, would stop at a 2-m distance and then perform the "Avoiding" maneuver according to the pre-programmed ROS environment algorithm before resuming movement. Table A.10 summarizes the characteristics of Sidewalk 2 at the University of Seoul, detailing the section length, types of obstacles, driving method, and avoidance points.

The third test section was a 70-m stretch of Sidewalk 3 at the University of Seoul, as shown in Fig. A.15. In this section, PM (Personal Mobility) devices and bicycles were selected as obstacles for testing. The robot sequentially detected PM devices and bicycles stopped 2 m before the obstacle and then performed an avoidance maneuver. Table A.11 summarizes the characteristics of the sidewalk test section, providing details on the test environment and obstacle avoidance process.

A.7 Detailed Scenario Results and Implications

This study aimed to develop a system that enables an autonomous robot to assess and avoid sidewalk damage and obstacles by applying an image-based algorithm. Key findings from this research include obstacle detection and avoidance, which were analyzed using Rosbag data and GPS data.

Figure A.16 presents the command code implemented in the ROS environment, which instructs the robot to "Avoiding" at a 2-m distance upon detecting a trained obstacle. When no obstacles or sidewalk damage are detected, the system is designed to allow the robot to continue normal driving without avoidance by issuing a "Continuing" command. The experimental results confirmed that in all three test sections, the robot successfully stopped at the 2-m mark and performed an avoidance maneuver, as implemented in the system.

Figure A.17 shows the rqt_graph, which visualizes the executed topic nodes in the Rosbag file. The executed topics indicate that image data and depth images from the camera, /vel (velocity data) from the IMU, and /fix latitude and longitude data from RTK-GPS can be received through communication.

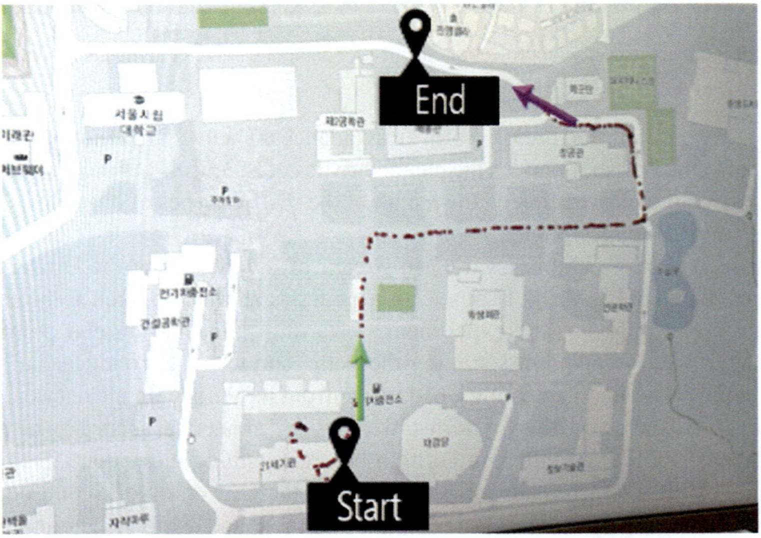

Fig. A.11 ROS environment—autonomous driving simulation at University of Seoul

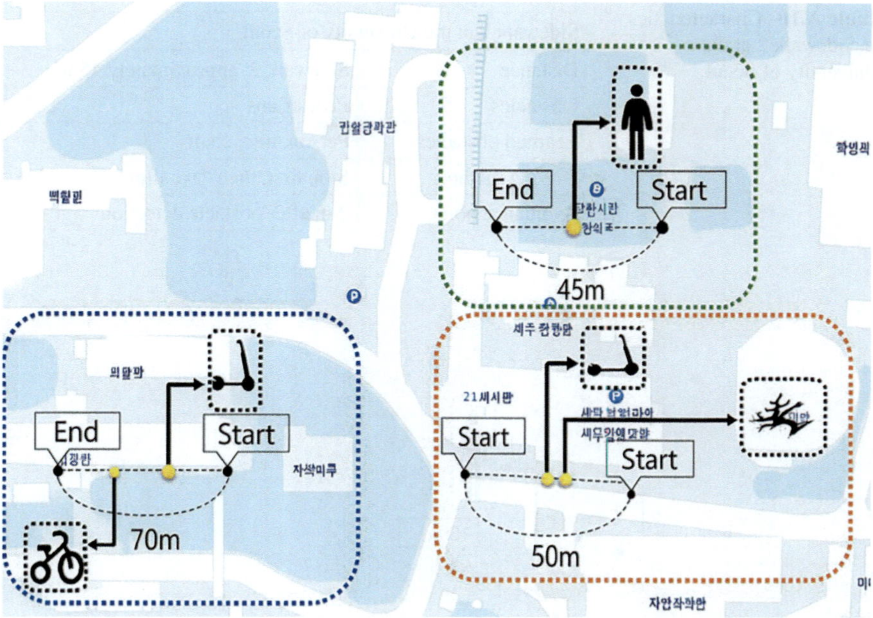

Fig. A.12 Sidewalk test section at University of Seoul

Table A.9 Characteristics of Sidewalk 1 at the University of Seoul

Sidewalk 1 at the University of Seoul	
Distance	Sidewalk 1: approximately 50 m
Obstacles	PM (personal mobility), surface damage
Learned obstacles	PM, Bin, Tree, Person, Break, Car
Driving method	Stop first, then "avoiding"
Avoidance point	2 m after obstacle detection

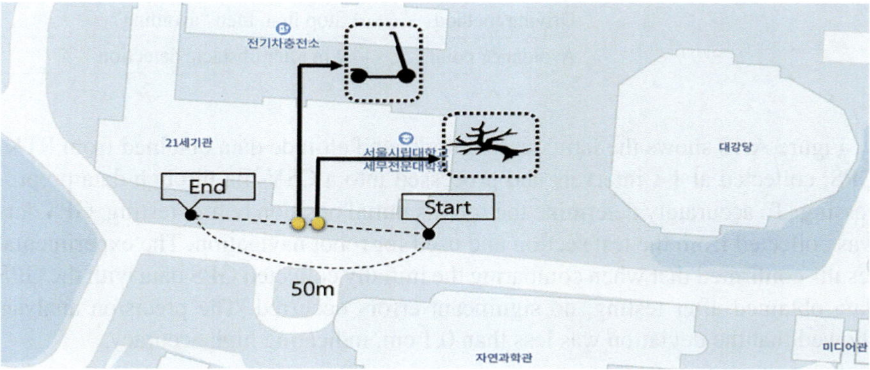

Fig. A.13 Sidewalk 1 at the University of Seoul

Table A.10 Characteristics of Sidewalk 2 at the University of Seoul

Sidewalk 2 at the University of Seoul	
Distance	Sidewalk 2: approximately 45 m
Obstacles	Pedestrians
Learned obstacles	Person, tree, chair
Driving method	Stop first, then "avoiding"
Avoidance point	2 m after obstacle detection

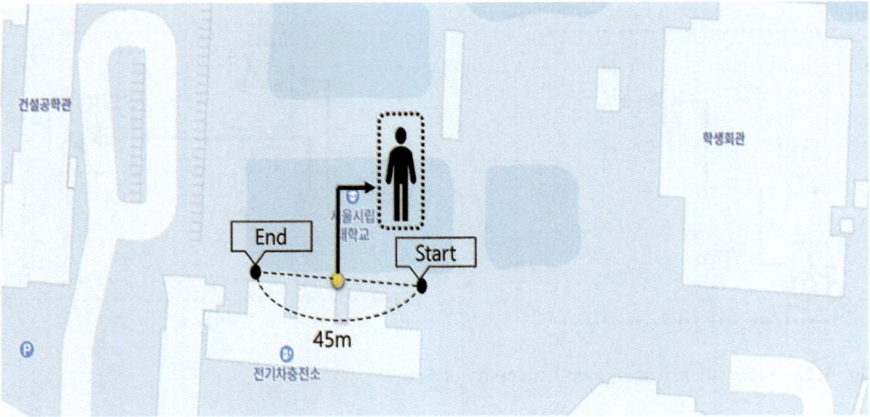

Fig. A.14 Sidewalk 2 at the University of Seoul

Table A.11 Characteristics of Sidewalk 2 at the University of Seoul

Sidewalk 3 at the University of Seoul	
Distance	Sidewalk 3: approximately 70 m
Obstacles	PM, Bicycle
Learned obstacles	PM, Tree, Bicycle
Driving method	Stop first, then "avoiding"
Avoidance point	2 m after obstacle detection

Figure A.18 shows the latitude, longitude, and altitude data obtained from RTK-GPS, collected at 1-s intervals and processed into a CSV file through data preprocessing. To accurately determine the robot's initial position before testing, GPS data was collected from the test section and used for robot navigation. The experimental results confirmed that when comparing the initially collected GPS data with the GPS data obtained after testing, no significant errors occurred. The precision analysis showed that the deviation was less than 0.1 cm, indicating high accuracy.

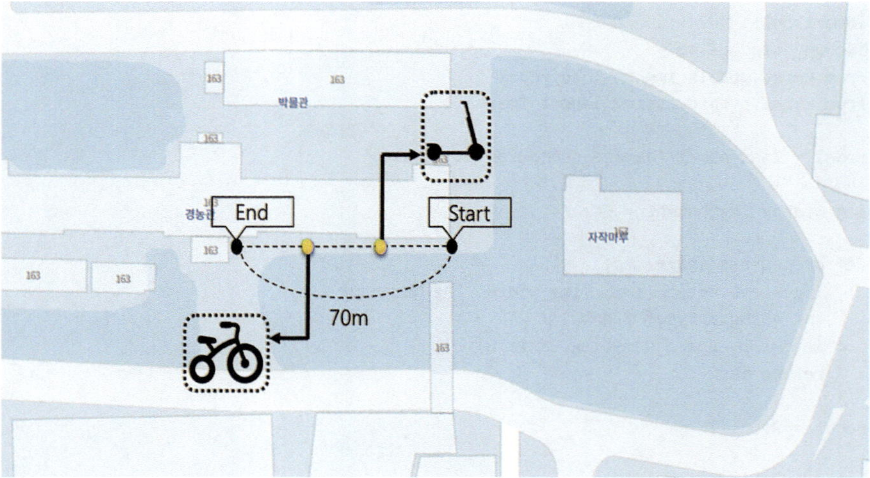

Fig. A.15 Sidewalk 3 at the University of Seoul

A.7.1 Sidewalk 1 (PM, Surface Damage)

The first experiment, conducted on Sidewalk 1 at the University of Seoul, evaluated the robot's ability to detect PM (Personal Mobility) devices and surface damage. The robot was configured to identify and avoid obstacles by applying a pre-trained algorithm. In addition, GPS data was used to allow the robot to autonomously drive in a straight path. Figure A.19 shows an image of GPS data extracted using Rosbag while navigating Sidewalk 1 in the ROS environment. Figure A.20 presents the preprocessed data showing the robot's avoidance path after being commanded via navigation to move to the endpoint, detecting and avoiding obstacles such as PM devices and surface damage. For comparison of GPS data, only GPS topics were extracted from the Rosbag, converted into a CSV file, and analyzed using the ArcGIS program. Figures A.21 and A.22 show image frames captured by the robot's depth camera during the experiment, integrated with the detection algorithm to assess performance. Figures A.23 and A.24 are mobile phone photos capturing the moments when the robot detects, evaluates, and avoids the obstacles. The results confirmed that the robot successfully recognized not only PM devices but also surrounding obstacles such as trees, pedestrians, and various levels of surface damage, along with their context. Figures A.25 and A.26 show images from the robot's point of view while detecting obstacles on Sidewalk 1. Figures A.27 and A.28 illustrate the robot's obstacle avoidance (PM) and autonomous navigation in action.

```
import cv2
import numpy as np
from keras.models import load_model
from keras.preprocessing import image

model = load_model('damage_detection_model.h5')

img_width, img_height = 64, 64

def preprocess_image(img):
    img = cv2.resize(img, (img_width, img_height))
    img = img.astype("float") / 255.0
    img = np.expand_dims(img, axis=0)
    return img

robot = Robot()

cap = cv2.VideoCapture(0)

while True:
    ret, frame = cap.read()

    if not ret:
        continue

    proc_img = preprocess_image(frame)
    predictions = model.predict(proc_img)

    if predictions[0][0] >= 0.5:
        print("Damage detected. Avoiding...")
        robot.avoid_damage()
    else:
        print("No damage detected. Continuing...")
        robot.continue_driving()

    cv2.putText(frame, f"Damage Prob: {predictions[0][0]:.2f}", (10, 30),
                cv2.FONT_HERSHEY_SIMPLEX, 0.9, (0, 255, 0), 2)
    cv2.imshow('Robot Camera View', frame)

    if cv2.waitKey(1) & 0xFF == ord('q'):
        break

cap.release()
cv2.destroyAllWindows()
```

Fig. A.16 System implementation code for assessing robot damage and obstacles

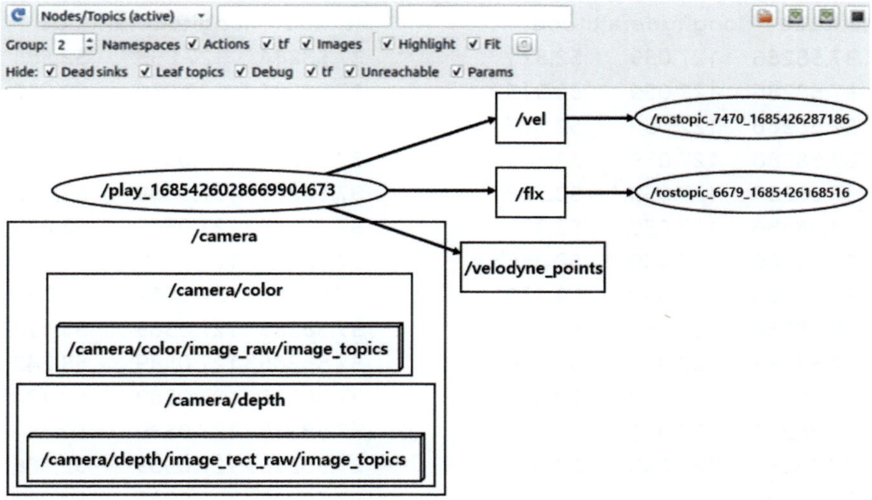

Fig. A.17 Rosbag topic nodes

A.7.2 Sidewalk 2 (Pedestrian)

The second experiment was conducted on Sidewalk 2 at the University of Seoul, focusing on the robot's ability to recognize pedestrians and surrounding conditions. Figure A.29 shows the GPS data extracted through Rosbag during the robot's test on Sidewalk 2. Figure A.30 displays preprocessed data from Rosbag, showing that the robot, after receiving a command via RTK-GPS to reach the end point, detected a pedestrian obstacle, performed an avoidance maneuver, and successfully reached the destination. Figures A.31 and A.32 show images from the robot's point of view as it navigated through Sidewalk 2 after training with obstacle data. Figures A.33 and A.34 are mobile phone photos capturing the robot as it performed avoidance behavior upon encountering a pedestrian. The experiment confirmed that the robot was able to accurately recognize not only people, but also sidewalk obstacles such as trees and chairs, as well as the overall surroundings. Figures A.35 and A.36 show the robot's view during the recognition of learned obstacles on Sidewalk 2. Figures A.37 and A.38 illustrate the robot performing obstacle avoidance and autonomous driving.

A.7.3 Sidewalk 3 (PM, Bicycle)

The third test section was conducted on Sidewalk 3 at the University of Seoul and evaluated the robot's ability to recognize PM (Personal Mobility) devices, bicycles, and surrounding conditions. Figure A.39 shows the GPS data extracted from ROS during the robot's test on Sidewalk 3. Figure A.40 presents GPS-based confirmation

latitude	longitude	altitude	latitude	longitude	altitude
37.58286	127.059	52.321	37.58284	127.059	52.588
37.58286	127.059	52.315	37.58284	127.059	52.552
37.58286	127.059	52.327	37.58285	127.0589	52.546
37.58286	127.059	52.341	37.58285	127.0589	52.667
37.58286	127.059	52.322	37.58285	127.0589	52.595
37.58286	127.059	52.335	37.58285	127.0589	52.873
37.58286	127.059	52.465	37.58285	127.0589	52.83
37.58286	127.059	52.555	37.58285	127.0589	52.91
37.58286	127.059	52.599	37.58285	127.0589	53.197
37.58286	127.059	52.766	37.58285	127.0589	53.242
37.58286	127.059	52.838	37.58285	127.0589	53.124
37.58286	127.059	52.926	37.58285	127.0589	52.841
37.58286	127.059	53.077	37.58285	127.0589	52.766
37.58286	127.059	53.125	37.58285	127.0589	52.681
37.58286	127.059	53.132	37.58285	127.0589	52.549
37.58286	127.059	53.06	37.58285	127.0589	52.546
37.58286	127.059	52.971	37.58285	127.0589	52.6
37.58286	127.059	52.949	37.58285	127.0588	52.499
37.58286	127.059	52.787	37.58285	127.0588	52.096
37.58286	127.059	52.757	37.58285	127.0588	51.752
37.58286	127.059	52.726	37.58285	127.0588	51.797
37.58286	127.059	52.712	37.58285	127.0588	51.893
37.58286	127.059	52.749	37.58285	127.0588	51.906

Fig. A.18 RTK-GPS data

that the robot, after receiving a command to reach the end point, avoided obstacles (PM and bicycle) at two locations and arrived at the destination. Figures A.41 and A.42 show image frames captured by the robot's camera combined with the detection algorithm to assess obstacle detection and avoidance performance. Figures A.43 and A.44 are mobile phone photos capturing the robot as it avoided trained PM and bicycle obstacles during the test. The experimental results confirmed that the robot was able to accurately recognize not only PM devices and bicycles, but also sidewalk obstacles such as trees, along with the surrounding environment. Figures A.45 and A.46 show the robot's point of view in recognizing learned obstacles on Sidewalk 3. Figures A.47, A.48, and A.49 illustrate a sequence of the robot performing obstacle avoidance (PM, bicycle) and autonomous driving.

```
header:
  seq: 85
  stamp:
    secs: 1685064123
    nsecs: 114773750
  frame_id: "gps"
status: 2
service: 1
latitude: 37.588586613333336
longitude: 127.02958821333334
altitude: 52.521
position_covariance: [26.625600000000002, 0.0, 0.0, 0.0, 26.625600000000002, 0.0, 0.0, 0.0, 426.0096000000003]
position_covariance_type: 1

header:
  seq: 87
  stamp:
    secs: 1685064125
    nsecs: 118254661
  frame_id: "gps"
status: 2
service: 1
latitude: 37.588586681666665
longitude: 127.02958829333333
altitude: 52.513000000000005
position_covariance: [26.625600000000002, 0.0, 0.0, 0.0, 26.625600000000002, 0.0, 0.0, 0.0, 426.0096000000003]
position_covariance_type: 1

header:
  seq: 88
  stamp:
    secs: 1685064126
    nsecs: 116723585
  frame_id: "gps"
status: 2
service: 1
latitude: 37.58858681666667
longitude: 127.029588295
altitude: 52.537
position_covariance: [26.625600000000002, 0.0, 0.0, 0.0, 26.625600000000002, 0.0, 0.0, 0.0, 426.0096000000003]
position_covariance_type: 1
```


Fig. A.19 GPS data extracted from ROS (Sidewalk 1)

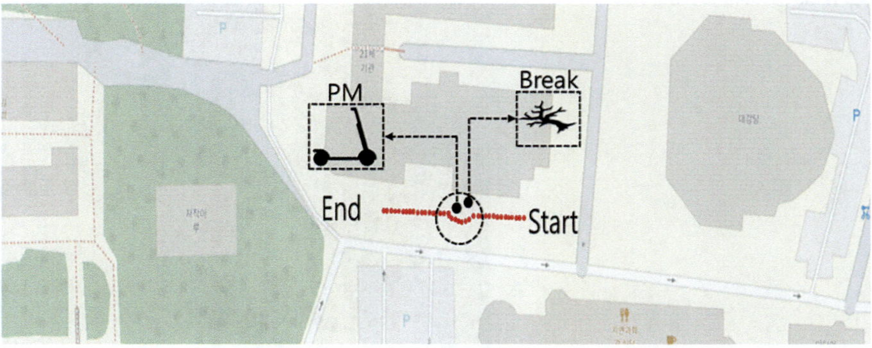

Fig. A.20 Post-processing of GPS coordinates after autonomous driving command

Fig. A.21 Obstacle detection (robot's point of view)

Fig. A.22 Surrounding obstacle detection

Fig. A.23 Robot driving

Fig. A.24 Robot obstacle avoidance

A.7.4 Detailed Scenario Results and Implications

For each section analyzed in the scenario, an accurate analysis was conducted by comparing the predicted values and actual values for obstacle and surface damage classification. The predicted results were extracted frame-by-frame from the corresponding raw data. By classifying the positive and negative outcomes for each detection result, accuracy rates were calculated and analyzed for each case. From this scenario analysis, a total of eight obstacle types were identified: Tree, Person, PM (Personal Mobility), Bin, Car, Chair, Light, and Bicycle. Using a confusion matrix, the probabilities between predicted and actual values were calculated as shown in Appendix Fig. A.50. Based on the eight detected classes, the Positive-Positive accuracy rate was found to be 96.92%.

An analysis was conducted to identify the causes of discrepancies between the predicted and actual values in the scenario results. Based on this analysis, it was found that improvements such as enhanced architectural design to better distinguish visually similar images, and better handling of shadows and distant objects, are necessary. Continuous refinement of the algorithm is essential to minimize misclassification of obstacles and surface damage in sidewalk environments.

In Fig. A.51, the actual class should be "Person", but it was incorrectly predicted as PM (Personal Mobility). As shown in the upper portion of Fig. A.51, the original image depicts a person riding a bicycle. In such a case, the system should have detected both "Person" and "Bicycle", even from a distance. However, due to the visual overlap and the bending posture of the person on the bicycle, the algorithm may have interpreted the scene as resembling a PM. This highlights the need for more detailed training on scenarios where a person is riding another vehicle, and the architecture should be extended to handle such composite forms.

In Fig. A.52, the actual class is "Car", but it was mistakenly predicted as "Person". This likely occurred due to shadows from nearby trees and the convex oval shape of the car, which could visually resemble a person from a distance. To address such

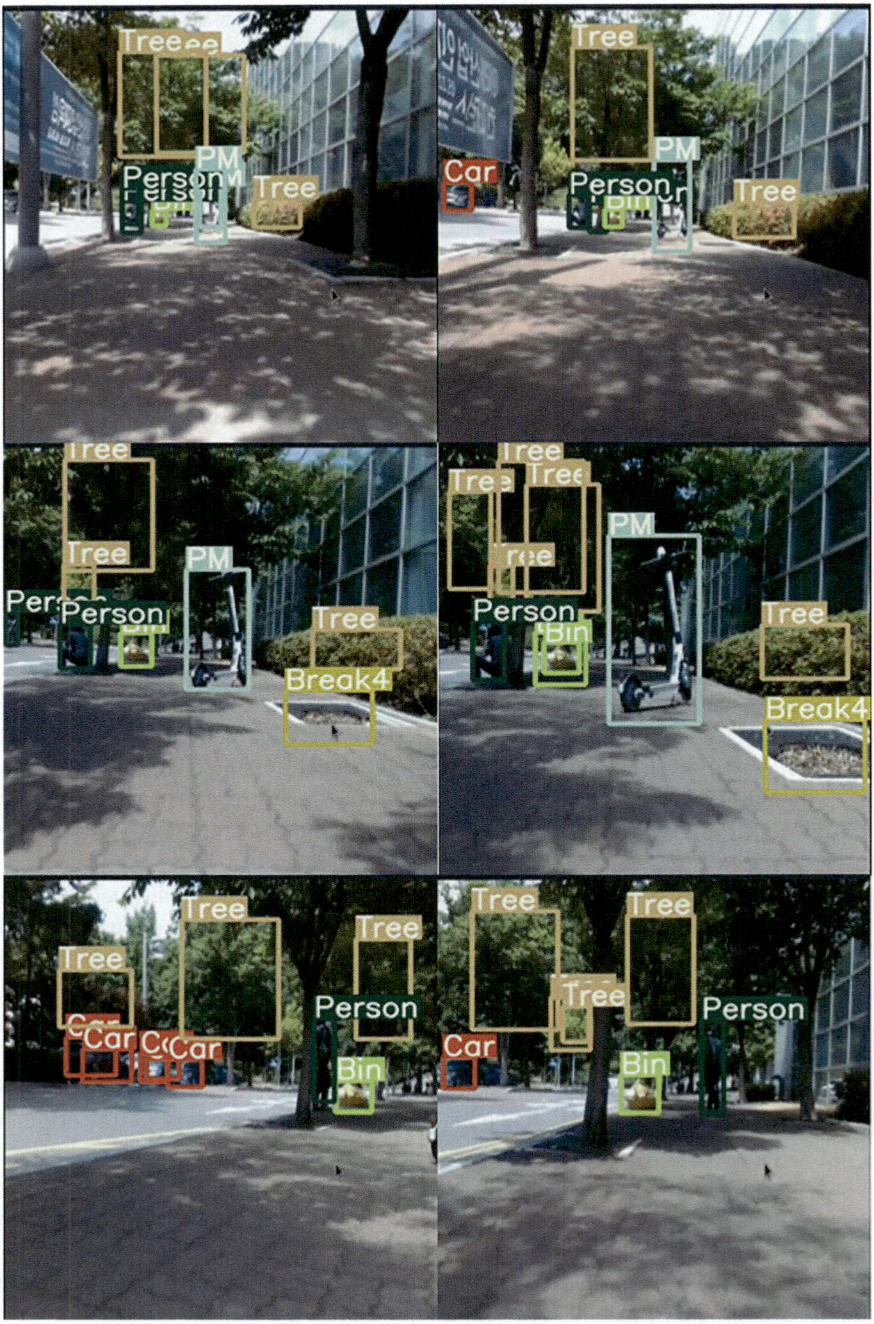

Fig. A.25 Obstacle detection on Sidewalk 1 (robot's point of view) (1)

Fig. A.26 Obstacle detection on Sidewalk 1 (robot's point of view) (2)

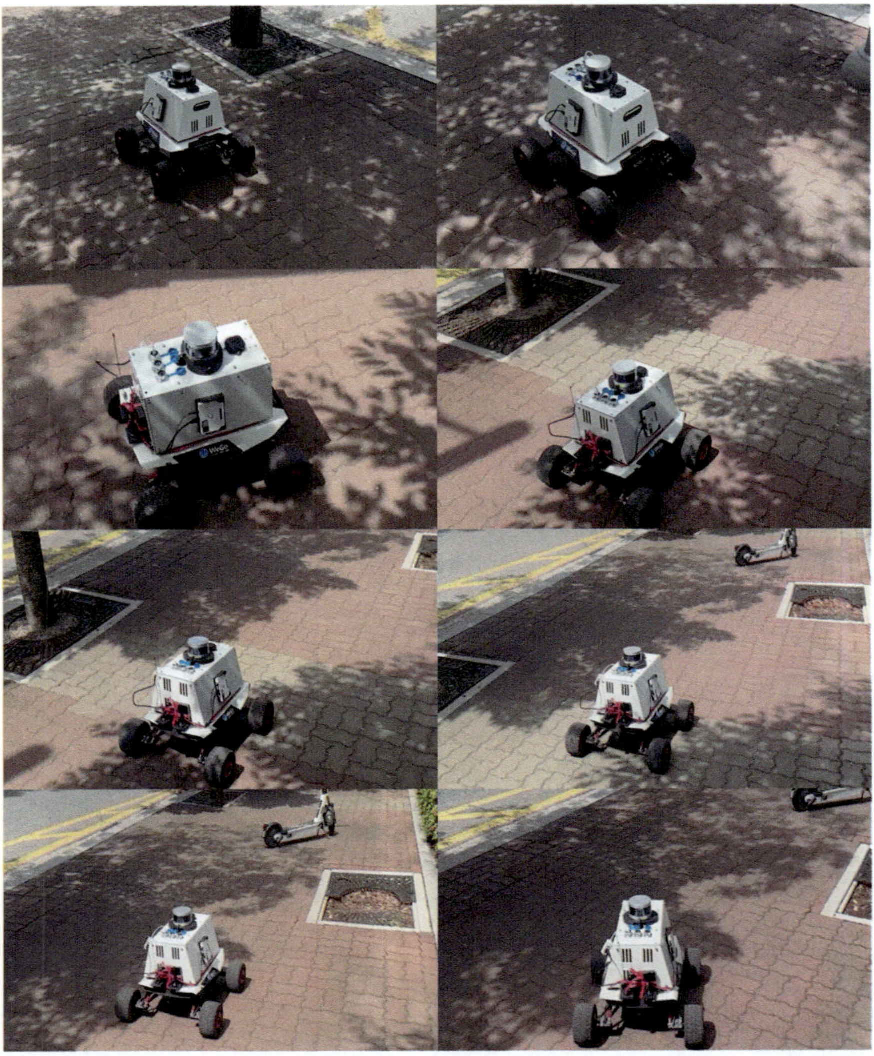

Fig. A.27 Obstacle (PM) avoidance and autonomous driving (1)

errors, further training on distant object detection and shadow conditions should be incorporated.

In Fig. A.53, the actual class is "Light", but the object was misclassified as PM and Tree. The elongated shape of the light pole was not clearly visible due to shadows and surrounding trees. The system appears to have confused the long vertical shape with that of a PM device. In another frame, the same object was recognized as a Tree. This is likely because the upper part of the object overlaps with nearby trees, while the lower part corresponds to the actual light pole. Due to shadowing in the

Fig. A.28 Obstacle (PM) avoidance and autonomous driving (2)

image, the system interpreted the entire area as a tree and applied the bounding box accordingly. Since the input is in 2D, the system lacks the depth perception that a human observer would use to distinguish that the light pole is behind the tree. To address this limitation, additional segmentation techniques or fine-tuned object separation strategies should be applied to enable clearer distinction between overlapping objects in 2D images.

```
header:
  seq: 46
  stamp:
    secs: 1685066708
    nsecs: 987481810
  frame_id: "gps"
status: 2
service: 1
latitude: 37.583608925
longitude: 127.05909153
altitude: 49.159
position_covariance: [27.457600000000003, 0.0, 0.0, 0.0, 27.457600000000003, 0.0, 0.0, 0.0, 439.32160000000005]
position_covariance_type: 1
---
header:
  seq: 47
  stamp:
    secs: 1685066709
    nsecs: 91104030
  frame_id: "gps"
status: 2
service: 1
latitude: 37.58361945
longitude: 127.05909186666666
altitude: 49.198
position_covariance: [13.838400000000002, 0.0, 0.0, 0.0, 13.838400000000002, 0.0, 0.0, 0.0, 221.41440000000003]
position_covariance_type: 1
---
header:
  seq: 48
  stamp:
    secs: 1685066710
    nsecs: 91387163
  frame_id: "gps"
status: 2
service: 1
latitude: 37.58362858333333
longitude: 127.05909158333334
altitude: 49.253
position_covariance: [13.838400000000002, 0.0, 0.0, 0.0, 13.838400000000002, 0.0, 0.0, 0.0, 221.41440000000003]
position_covariance_type: 1
```


Fig. A.29 GPS data extracted from ROS (Sidewalk 2)

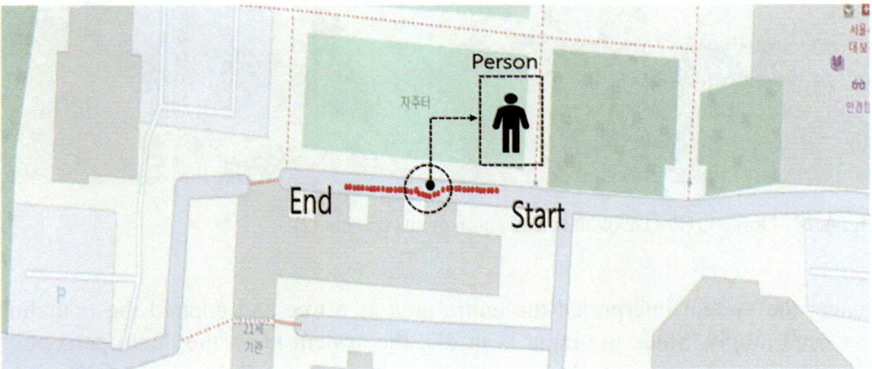

Fig. A.30 Post-processing of GPS coordinates after autonomous driving command

Fig. A.31 Surrounding obstacle detection

Fig. A.32 Obstacle detection

Fig. A.33 Robot driving

Fig. A.34 Robot obstacle (pedestrian) avoidance

In this scenario, one class of surface damage was detected. The model identified patterns that differed from the typical sidewalk blocks and assessed the degree of damage to generate results, as shown in Fig. A.54. However, previous results had been derived using video data recorded from a fixed direction, without significant changes in camera orientation, as shown in Fig. A.55. During this robot test, the camera angle was lowered compared to the human-eye-level videos used before, and the direction of movement also changed. Despite these changes, the model achieved an accuracy of 91.03%for damage detection in a single result. To maintain high accuracy in damage detection even when camera angles and movement directions vary, it is necessary to incorporate additional complex factors into the model to further enhance its performance.

A.8 Conclusion

In order to achieve efficient robot navigation on sidewalks, a pedestrian environment assessment system was first developed and combined with an autonomous navigation algorithm, and tests were conducted on sidewalk sections at the University of Seoul to evaluate obstacle detection and surface damage on sidewalks. The robot system performed safe navigation by normally avoiding trained obstacles, and the driving path was verified through GPS data. Based on the research results, it is expected that a system capable of responding to subdivided levels of sidewalk damage and to damage sizes can be established in the future.

The trained model was integrated into the control system of the autonomous robot so that the robot could process the model's output and make real-time decisions based on that output. This system continuously captures images in the robot environment, preprocesses those images, and feeds them into the trained algorithm. The model then

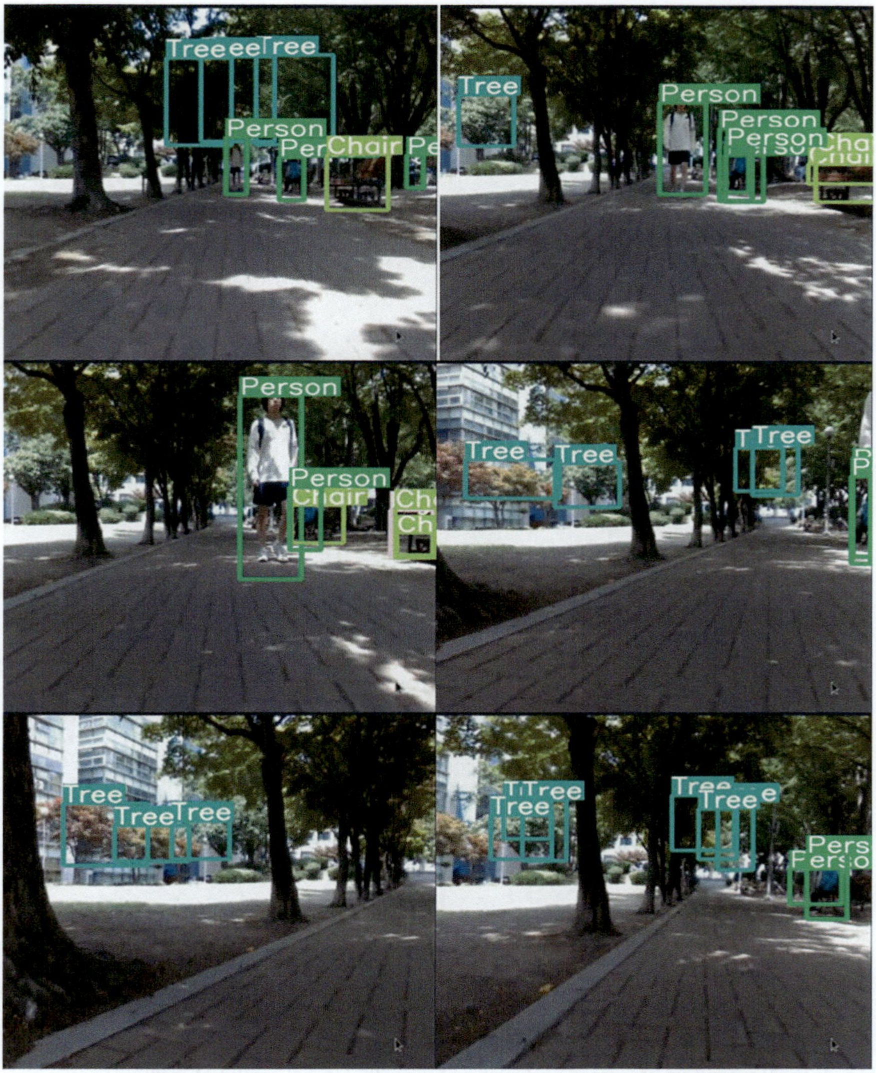

Fig. A.35 Recognition of learned obstacles on Sidewalk 2 (robot's point of view) (1)

analyzes each image and predicts the presence or absence of obstacles. Depending on the model's output, the robot's control system makes decisions. If an obstacle is detected with high probability, the robot is programmed to stop or change its path to effectively avoid the obstacle. If no obstacle is expected, the robot continues forward without change. To test the robustness of the system, experiments were conducted in various real environments. The system demonstrated excellent performance by avoiding various obstacles such as trash bins, uneven surfaces, and pedestrians.

Fig. A.36 Recognition of learned obstacles on Sidewalk 2 (robot's point of view) (2)

In addition, it is expected that optimal navigation across more sidewalk sections will become possible through continuous learning and updating of various types of obstacles that may occur on sidewalks. Future research will include developing a system that integrates sidewalk and roadway networks in parallel and conducting studies that fully reflect the characteristics of sidewalks by considering path planning algorithms based on human trajectories, algorithms that determine pass ability based on hardware size and surface damage, and path planning algorithms that consider slope. Through this, it will be possible to achieve optimal movement of autonomous robots on sidewalks efficiently.

Fig. A.37 Obstacle (pedestrian) avoidance and autonomous driving (1)

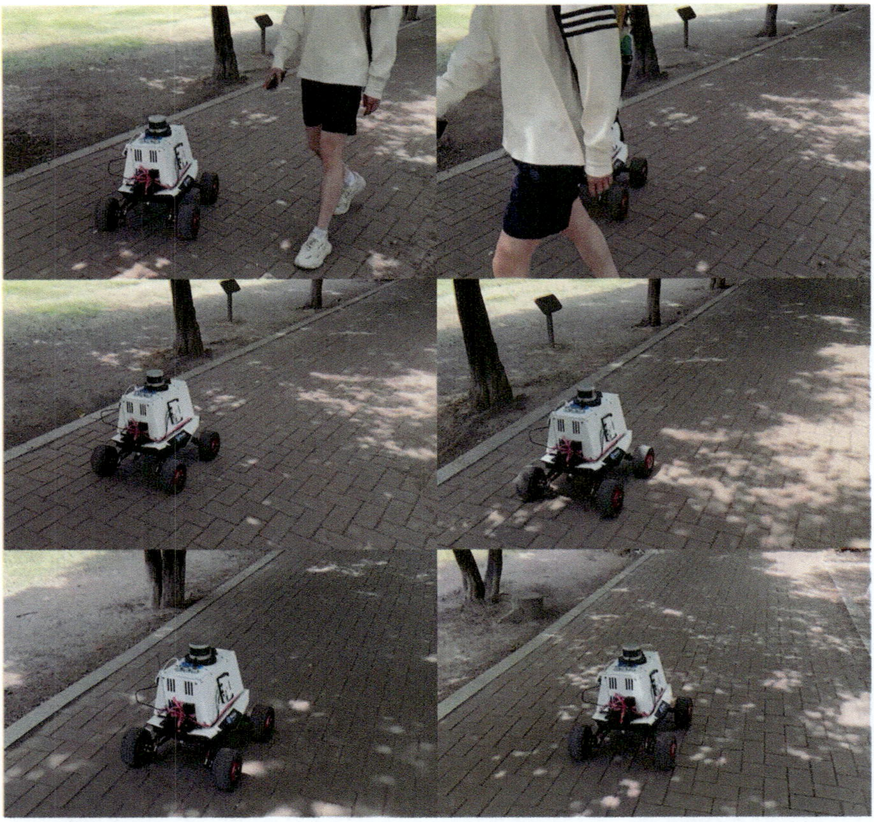

Fig. A.38 Obstacle (pedestrian) avoidance and autonomous driving (2)

```
header:
  seq: 106
  stamp:
    secs: 1685081164
    nsecs: 107550859
  frame_id: "gps"
status:
  status: 1
  service: 1
latitude: 37.583793203333336
longitude: 127.05344615
altitude: 72.968
position_covariance: [0.008464000000000003, 0.0, 0.0, 0.0, 0.008464000000000003, 0.0, 0.0, 0.0, 0.13542400000000004]
position_covariance_type: 1
---
header:
  seq: 110
  stamp:
    secs: 1685081168
    nsecs: 107889175
  frame_id: "gps"
status:
  status: 1
  service: 1
latitude: 37.583781028333334
longitude: 127.05346906166666
altitude: 71.318
position_covariance: [0.008464000000000003, 0.0, 0.0, 0.0, 0.008464000000000003, 0.0, 0.0, 0.0, 0.13542400000000004]
position_covariance_type: 1
---
header:
  seq: 112
  stamp:
    secs: 1685081170
    nsecs: 109625816
  frame_id: "gps"
staus:
  status: 1
  service: 1
latitude: 37.58378729166667
longitude: 127.05349240833333
altitude: 67.67500000000001
position_covariance: [0.009216, 0.0, 0.0, 0.0, 0.009216, 0.0, 0.0, 0.0, 0.147456]
position_covariance_type: 1
```


Fig. A.39 GPS data extracted from ROS (Sidewalk 3)

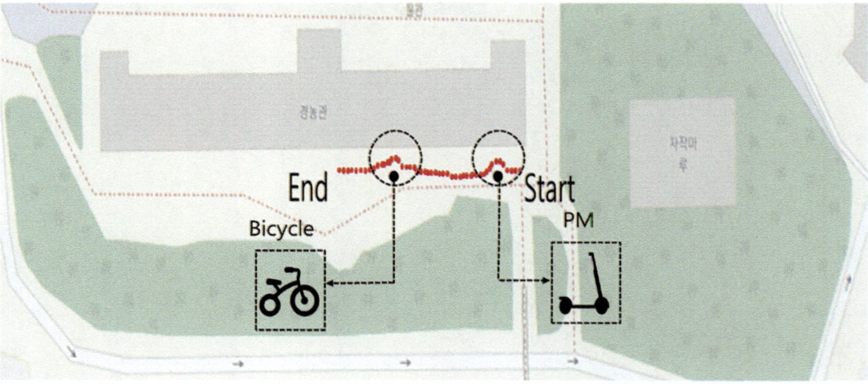

Fig. A.40 Post-processing of GPS coordinates after autonomous driving command

Fig. A.41 Obstacle detection (PM)

Fig. A.42 Obstacle detection (bicycle)

Fig. A.43 PM avoidance

Fig. A.44 Bicycle avoidance

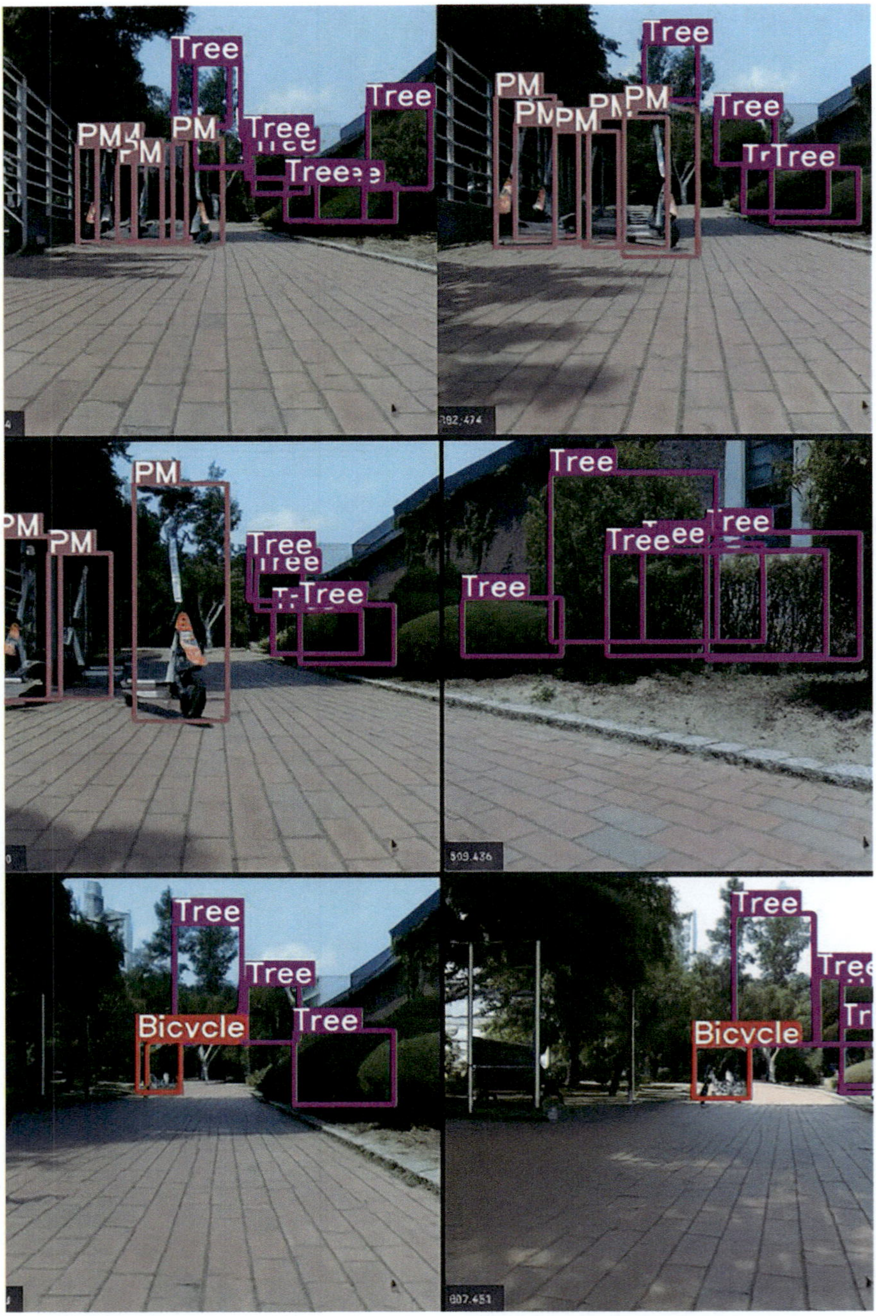

Fig. A.45 Recognition of learned obstacles on Sidewalk 3 (robot's point of view) (1)

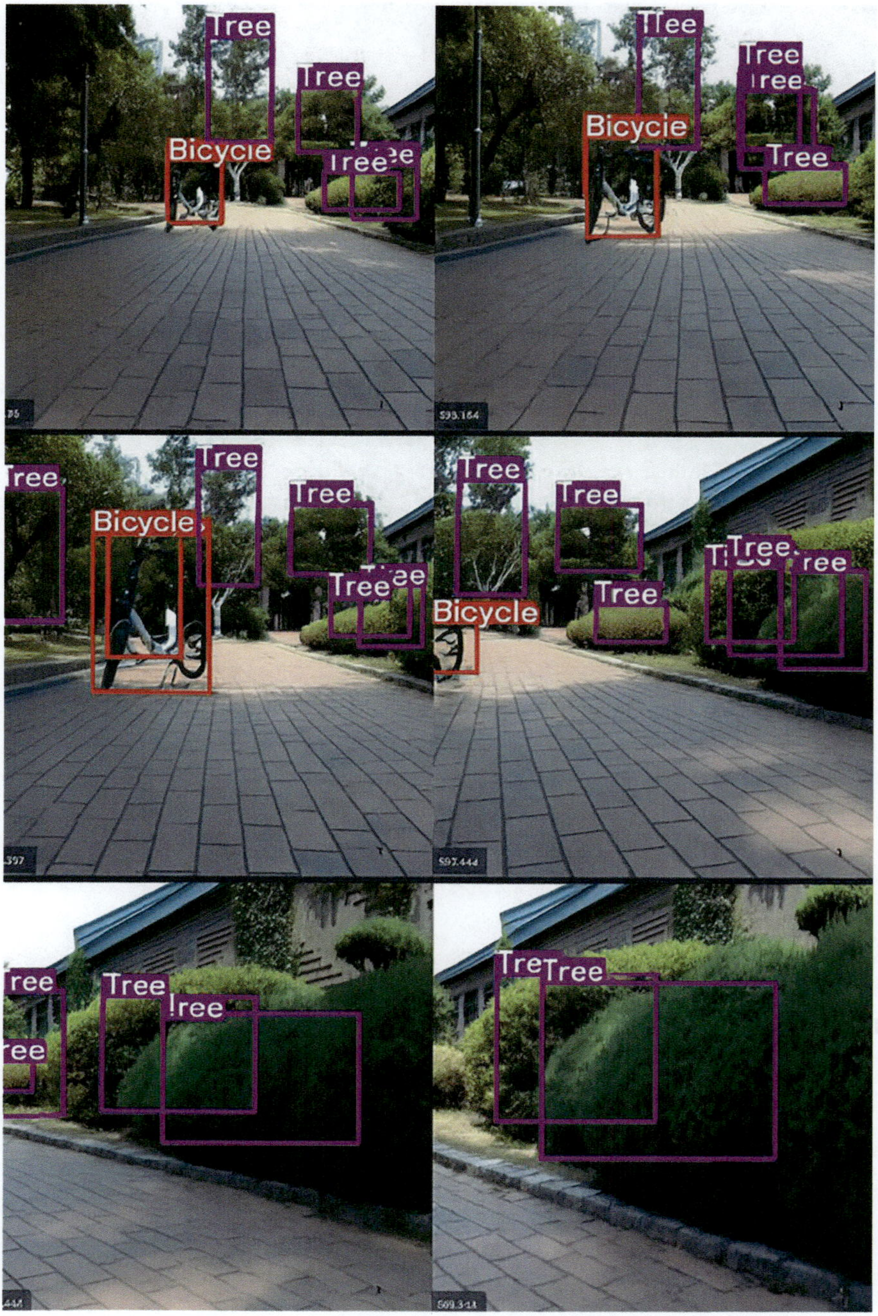

Fig. A.46 Recognition of learned obstacles on Sidewalk 3 (robot's point of view) (2)

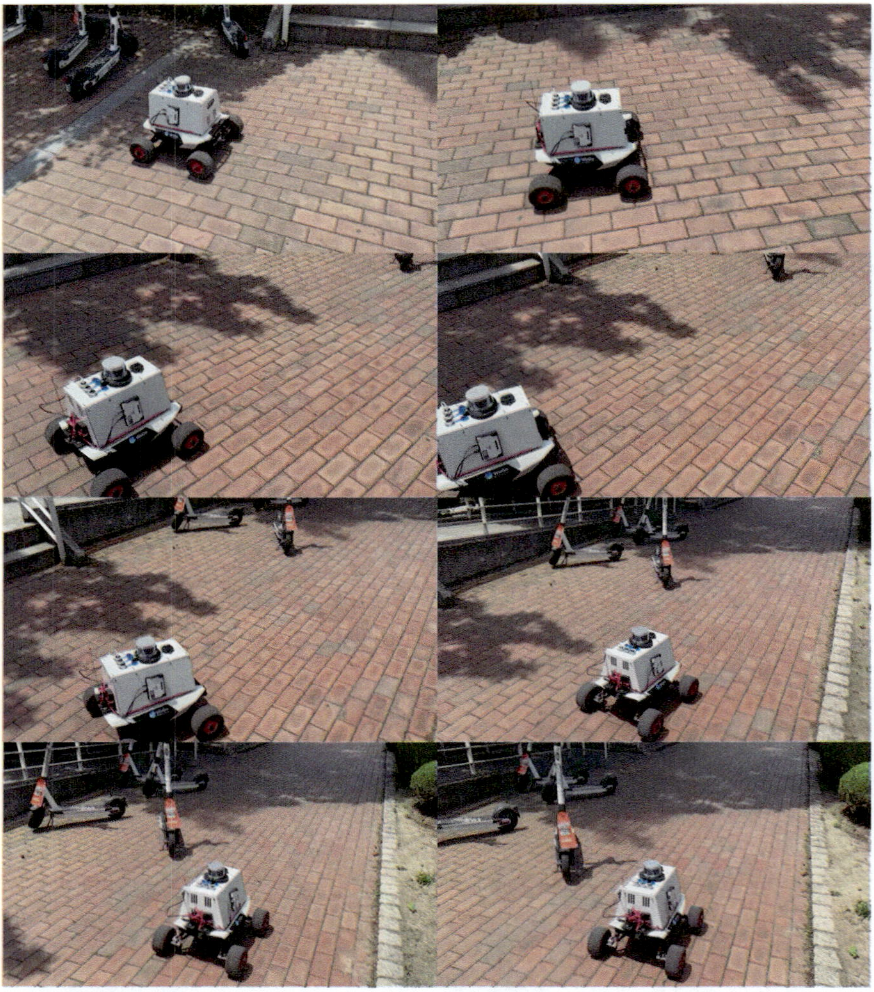

Fig. A.47 Sequence of obstacle (PM, bicycle) avoidance and autonomous driving (1)

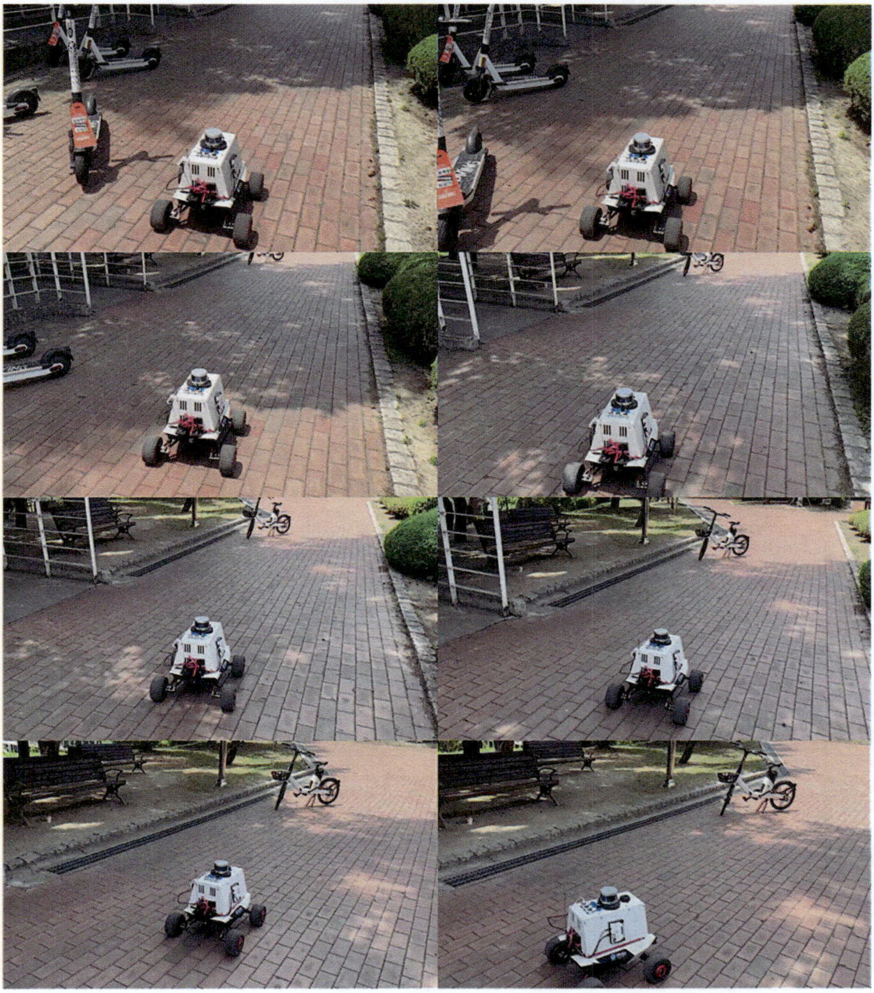

Fig. A.48 Sequence of obstacle (PM, bicycle) avoidance and autonomous driving (2)

Fig. A.49 Sequence of obstacle (PM, bicycle) avoidance and autonomous driving (3)

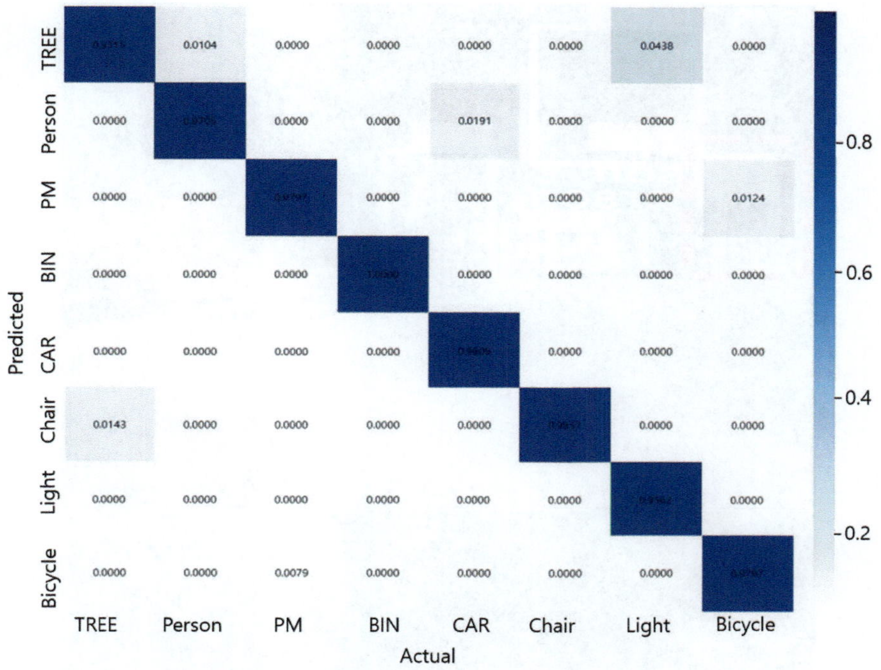

Fig. A.50 Obstacle classification results using the robot

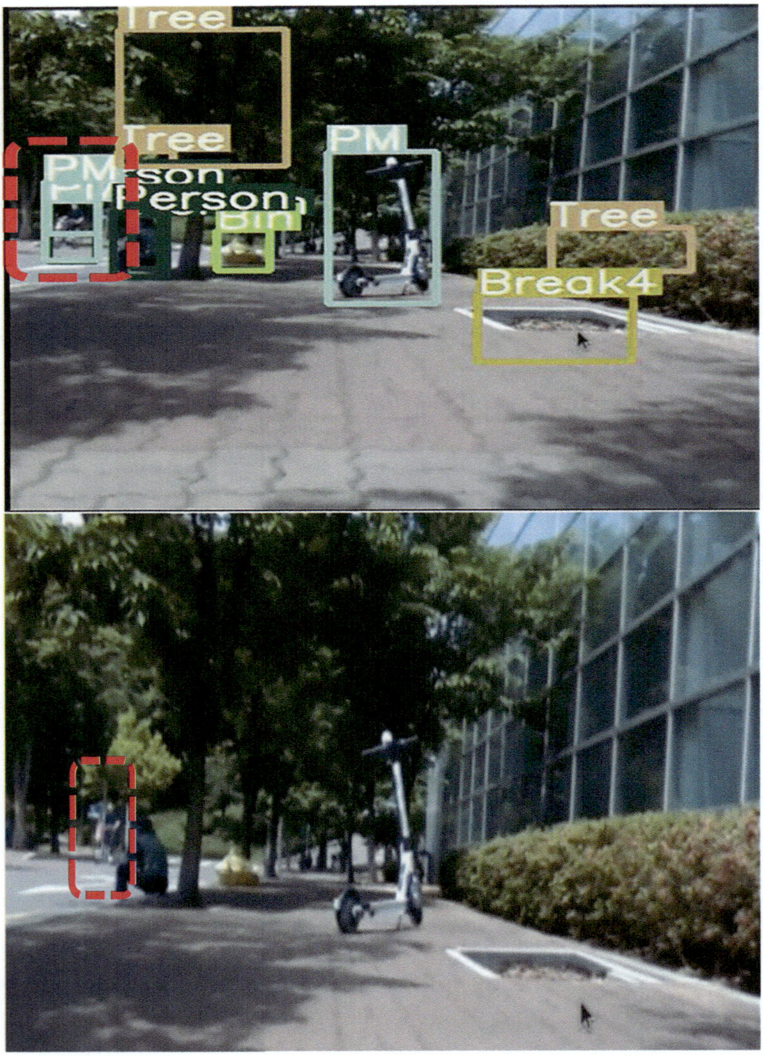

Fig. A.51 Error example—actual: person, predicted: PM (top: detection result, bottom: original image)

Fig. A.52 Error example—actual: car, predicted: person (top: detection result, bottom: original image)

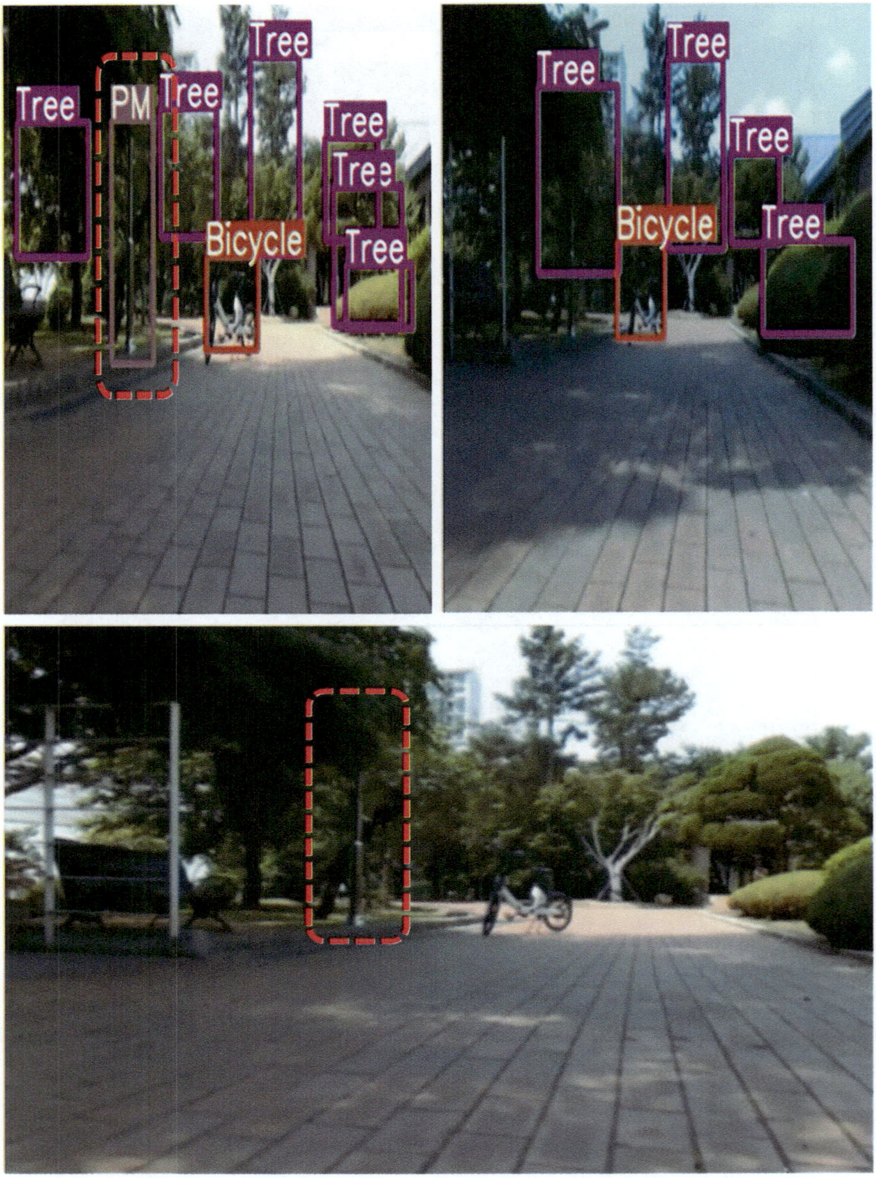

Fig. A.53 Error example—actual: light, predicted: tree (top: detection result, bottom: original image)

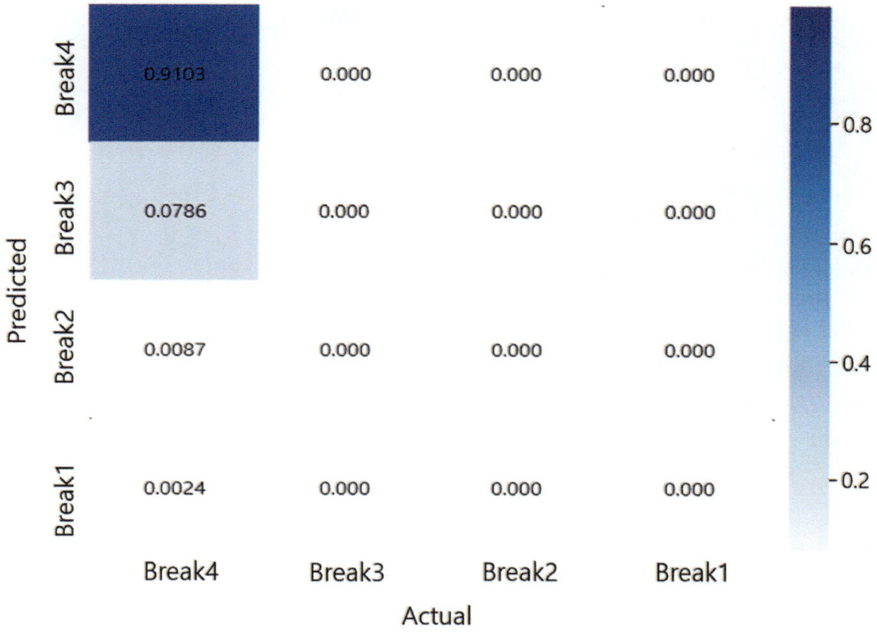

Fig. A.54 Surface damage classification results using the robot

Fig. A.55 Accuracy and variation in damage detection according to directional change

GPSR Compliance

The European Union's (EU) General Product Safety Regulation (GPSR) is a set of rules that requires consumer products to be safe and our obligations to ensure this.

If you have any concerns about our products, you can contact us on ProductSafety@springernature.com

In case Publisher is established outside the EU, the EU authorized representative is:

Springer Nature Customer Service Center GmbH
Europaplatz 3
69115 Heidelberg, Germany

Batch number: 10704307

Printed by Printforce, the Netherlands